Emerging Plant Communication in Contemporary Societies

Communication, Environment, Science and Society Set

coordinated by
Andrea Catellani and
Céline Pascual Espuny

Volume 5

Emerging Plant Communication in Contemporary Societies

Beyond the Human World

Edited by
Natacha Souillard
Amélie Coulbaut-Lazzarini
Céline Pascual Espuny

ISTE

WILEY

First published 2026 in Great Britain and the United States by ISTE Ltd and John Wiley & Sons, Inc.

ISTE Ltd
27-37 St George's Road
London SW19 4EU
UK

www.iste.co.uk

John Wiley & Sons, Inc.
111 River Street
Hoboken, NJ 07030
USA

www.wiley.com

Library of Congress Control Number: 2026932680

British Library Cataloguing-in-Publication Data
A CIP record for this book is available from the British Library
ISBN 978-1-83669-020-7

The manufacturer's authorized representative according to the EU General Product Safety Regulation is Wiley-VCH GmbH, Boschstr. 12, 69469 Weinheim, Germany, e-mail: Product_Safety@wiley.com.

Contents

Introduction

Emerging Communication on Plants in Contemporary Societies: Beyond the Human World

From actively rewilding spaces, and allowing areas to return to a semi-wild state through non-intervention (especially during the Covid-19 pandemic), to viewing the natural environment as a provider of services, the ways in which we relate to the plant world are shifting. We are becoming more aware of the concrete impacts of global warming, and the major tipping points fast approaching which could cause systemic meltdown, as is demonstrated by the series of COP (Conference of the Parties) climate talks (Barraud et al. 2019). For example, 2021 was an important year for biodiversity. Instituted at the Earth Summit in Rio de Janeiro in 1992 by the Convention on Biological Diversity, followed by 2010's Nagoya Protocol on Biodiversity, there has been a significant rise in awareness of the importance of this environmental issue, but it is still not a truth universally acknowledged.

Every year, the IUCN's Red List of Threatened Species is published as a barometer regarding the situation, but there is no consensus as to the required methods of conservation, protection or defense; proof of this can be seen in the imprisonment of Paul Watson – founder of the NGO Sea Shepherd – without trial, in 2024, mandated by states who refuse to recognize these international protection measures.

Introduction written by Céline PASCUAL ESPUNY, Amélie COULBAUT-LAZZARINI and Natacha SOUILLARD.

More generally, public bodies are gradually becoming increasingly aware of the living world – sometimes overtly and visibly; sometimes discreetly, behind the scenes. Certain struggles have become iconic, and some of the symbols associated with those struggles have attracted a great deal of media attention. For example, polar bears, orangutans and large tigers have aroused immediate public sympathy for their plight, and have been held up as figureheads by NGOs fighting much broader phenomena, such as deforestation and the melting of the ice pack. Hence, the agenda and public attention mean that our opinion is polarized in relation to certain species – either under threat of extinction, or posing a threat (e.g. because they are invasive) – in an episodic manner, before moving onto other crises deemed worthy of coverage. These nonhuman "voices", then, are given only sparse and isolated doses of attention, and they may even be absent from mediation processes because they are "invisible" to us (Wandersee and Schussler 2001; Zhong Mengual and Morizot 2018; Batke et al. 2020; Morizot 2020). Nevertheless, they contribute to a growing awareness and understanding, which are becoming normalized by a parallel legislative fight, largely waged by animal welfare associations.

In addition to the animal kingdom, there is another form of life which is less conspicuous, less well covered by the media, but sometimes also closer to home and therefore easier to relate to. In looking at plants and their plight, we are encouraged to give more consideration to the living world, as we see climate disasters happening before our very eyes – for instance, the (mega) wildfires which ravage the world's forests every year, and heatwaves which stifle urban life. The plant kingdom is being given renewed consideration, and plants are being deployed in public spaces. While media coverage of plant life is less vociferous, it is nevertheless present (Catellani et al. 2019, 2023). Our relationship with plants, algae, mosses, fungi, trees, flowers and forests, for example, offers numerous fields for research. Beyond our relationship with iconic greenery, we are also seeing shifts in our relationships with light, food, fertilization, shade and color. From gardens to enchanted forests, from potting shed to plate, from jungle to our urban sycamores, from parks to wildlands, our relationship with time is changing, adapting to the rhythm of the seasons and to growth times which far surpass human lifespans. Our relationship with space is also shifting: from microscopic sequencing of the pharmacopoeia, through the greening of urban spaces, to macroeconomic management of massive agricultural operations. Thus, thinking about the plant world has become part of a new way of thinking about our relationship with the world as a whole.

Research in biology and ecology leads us to believe that plants should be perceived in a new light – notably as sensitive beings, as well as ones which are intelligent, or even conscious (Calvo and Keijzer 2009; Gagliano 2012; Tassin 2016; Trewavas 2016; Chamovitz 2017; Mancuso and Viola (2018); Calvo et al. 2020; Parise et al. 2020). Thinking about our communication on the subject of the plant world thus leads us to reflect upon and analyze our relationship with a primordial fabric of interconnected ecosystems, which is important in symbolic, environmental, economic and cultural terms. Such is the purpose of this book, which follows in the footsteps of the conference, "Au-delà du monde des humains, communication végétale émergente dans l'espace public" (Beyond the Human World: Emerging Communication on Plants in the Public Space), organized by IMSIC[1], and the research group "Communication, Environnement, Sciences et Société" (Communication, Environment, Science and Society) at the SFSIC[2], held in Aix-en-Provence on May 25, 2022. From the point of view of communication studies, we believe various areas for examination emerged from the discussion – methodological and thematic – so this multi-authored book is divided into five parts:

– Part 1: semiotic approaches to the diversity of forms of discourse about and in the plant kingdom.

– Part 2: co-construction of a new relationship with the living world and environmental education.

– Part 3: art and mediation of the plant world.

– Part 4: sensory aspects and narratives surrounding human–nonhuman relationships.

– Part 5: communication on plants as a means of transforming organizations' communications.

These parts run the gamut of the approaches used in ICS, such as the processes of construction and structuring of meaning and shared frames of reference, social and socio-discursive relations, influence phenomena, organizational identities and the processes of transmission, mediation, narration and narrative creation. The chapters of this book are essentially part of a qualitative perspective and, for most, an exploratory approach. Each chapter is briefly presented below.

1 Institut méditerranéen des sciences de l'information et de la communication – Mediterranean Institute of Information and Communication Sciences.

2 Société française des sciences de l'information et de la communication – French Society of Information and Communication Sciences.

In Chapter 1, semiotic issues are examined in an infocommunications approach which connects us to the world of plants. Andrea Catellani presents a research project on the way in which plant organisms – in particular, forests – are present in organizational communications, in the specific case of organizations devoted to promoting reforestation. He offers a multimodal semiotic analysis of the verbal and visual aspects of the strategic communication of certain organizations who collect funding from private individuals and businesses to support reforestation projects the world over. The author focuses on the following question: How are plant beings – in particular, the collective entity known as a "forest" – deployed in the multimodal promotional and persuasive discourse put out by these organizations? In what way is the forest itself part of the discourse and images intended to spur action aimed at preserving and rebuilding that forest?

Taking a firm epistemological approach to semiotics, in Chapter 2, Nicola Zengiaro sets out an ecology of the relations between the organic and non-organic worlds, in a shift from phyto-semiotics to physio-semiotics. According to the author, semiosis can actually be identified with life (Sebeok 2001). This means that not only do animals have the capacity to communicate; so too do plants. The author refers to the birth, in the 1980s, of a new field in semiotics: phytosemiotics, which looks at semiotic processes and information exchanges between a plant and its environment. Then, he offers a more modern interpretation, which allows us to rethink the nature of relations between plants and their environment, notably drawing upon the concept of "sensification", borrowed from the philosopher Coccia. In other words, from photosynthesis to absorption of minerals, it is apparent that plants can exchange, understand and translate signs of meaning derived from inorganic matter (Prodi 2021). Thus, everything is an expression of the ontology in which the living being takes shape and comes into existence (Descola 2005).

Chapter 3 begins Part 2. Using a semiotic and linguistic analysis of the media content created by a Quebec influencer, Erica Lippert uncovers the overlaps and paradoxes in tourist communication relating to the natural environment. By examining the main modalities of tourist discourse "and the iconic and discursive imaginaries of nature stemming from it", she highlights the fact that the relationship with this "wild" environment tends to become a consumerist relationship, based in part on the desire to find a phantasmagorical and instrumentalized form of authenticity. Thus, promotion of natural environments may endanger them by leading excessive numbers of people to visit.

In Chapter 4, Amélie Coulbaut-Lazzarini examines the question of connection between humans and plants. In the author's view, plants are integral to the structure of our relationship with the world, in the form of images or texts contributed by actors from various backgrounds, such as the photographer Yann Arthus-Bertrand and the forester Peter Wohlleben (Wohlleben 2017). The question becomes one of public interest when public and regional communications seize on these subjects and reflect on a shift in our relationship with the world.

In this context, we ask, to what extent does regional bodies' such as PNRs[3] or local authorities' support for programs including artistic, corporal aspects in connection with landscapes, show a trend toward plant life being taken into account when thinking about humans' relationship with the world? Coulbaut-Lazzarini focuses on the way in which these programs frame the relationship with plants from a perspective that is not based solely on a utilitarian logic, whereby the natural world is a resource for humans to exploit, but instead one of communication, of connection, of "diplomatic cohabitation" (Morizot 2017) with the rest of the living world.

Writing about visiting abandoned ruins as a form of environmental education, Nathanaël Wadbled adopts the perspective of critical heritage studies (Harrison and Sterling 2020). In Chapter 5, he considers the practices of heritage and urban exploration (urbex) from an environmental standpoint, viewing these ruins as "hybrid" objects which are both cultural and natural. The question then arises of how to present these spaces in the media, and the role played in this process by a guide. Beyond the spaces themselves, visitors are led to become aware of the biodiversity developing in human buildings, which is often not perceived in the urban environment (Arnould et al. 2011). Thus, environmental education requires not only explanatory discourse, but an experience which gives primacy to the sensory, to the indeterminate and to everything in its proper context.

To open Part 3 concerning the art and mediation of the plant world, in Chapter 6, Natacha Cyrulnik and Marine Brun-Franzetti look at the representation of nature in Native American documentary filmmaking. They explain how, in such movies, plants are shown to reveal the catastrophic

3 Parcs naturels régionaux (Regional Natural Parks).

consequences of colonization on autochthonous peoples' ways of living and identities, and also on the environment. The plant world – the non-human – can thus have just as important a place as humans in the creation of a narrative surrounding the land, and we see the deep-rooted interdependence between human beings and their environment. As a tool for raising awareness and a spur to drive people to action, the representation of the plant world is constructed to offer spectators a sensory experience, inviting them to rethink the way in which they view nature and the human/nonhuman relations woven into it.

How can the combination of art and science contribute to the overhaul of scientific mediations? To what extent do artistic representations of physical and living phenomena constitute invitations to comprehend the contemporary environmental challenges? Through a study of painted landscapes, Zhong Mengual and Morizot (2018) explain that our relationship with the living world and natural environments stems from the dichotomy between art and science, which drives us to a "crisis of sensitivity". This stance both implies and explains our inability to interpret our environment. What we need to do is relearn how to produce processes of interpretation, given that the culture of landscape painting has always taken us further away from an understanding of the living world. It is necessary to re-establish the practices of engaging with and interpreting the world around us (Jeanneret 1994; Haraway 2009). In Chapter 7, Marie-Caroline Heïd, Eva Sandri, Sarah Labelle, Emma Laurent and Valérie Méliani analyze the way in which the sculpture *Immersion, Augmented Nature* is part of this recovery of our ability to interpret the landscape. Created by artist Anne-Lise Koehler, this papier-mâché sculpture was made using recovered paper (from bric-a-brac or books destined for the dump). The goal here is to re-examine the relationship between the artistic and scientific aspects in the installation *Immersion, Augmented Nature*, through a sociotechnical device (Peeters and Charlier 1999) as an arrangement of human, material and imagined elements – a "private space" in which "something" can happen.

In Chapter 8, Philippe Hert examines the issue of sensoriality and narratives in human/nonhuman relations. He leads us through an ethnographical study, based on autobiographical accounts (some of which have received widespread media attention) connected to the use of a beverage based on two plants, commonly known as ayahuasca. Here, the author focuses on issues linked to the "translation" of these experiences, and how they should lead us to rethink human/nature relations, other than as widely sharing. Westerners' high level of interest in shamanism involving plants, the quest to find "authentic" experiences and the exoticism which it carries, have given rise to a tourist industry (Amselle 2013), relayed by the media. The author raises the question of "dialog" with plants, drawing on numerous examples found on YouTube. While it is true that we have a certain horizon of expectations, notably pre-set for us by the media (Lombardi 2021), accounts of experiences tell of encounters from a point of view other than that of a world which is difficult, or impossible, to reconstruct. Can we then speak of meeting with a plant?

Chapter 9 takes us on a journey of taste and heritage: producing, selling and drinking so-called living wine are stages in a conversation with non-humans. What happens when we move from a wine made with what might be termed "conventional" practices to one made with more sustainable practices, be it in the context of organic viticulture, biodynamic wine or the very recently developed "Natural Method Wines"? According to the author, everything shifts when we alter the boundaries of the naturalistic paradigm. If we start to look at the way in which meaning circulates among non-humans, plants and animals (Kohn 2017), the way in which it circulates between rural workers and the plants they care for (Kazic 2022), and the way in which these "natural" winegrowers forge discernible connections (Pineau 2019), we begin to be better equipped to describe what happens positively in the work of these winegrowers. Using the semiotics of cooperative transactions developed by Zacklad (2020), Pierre Beslay analyzes this relationship in terms of a conversation with the living world, from working the vine through the vinification process to eventual bottling.

The last two chapters form the fifth and final part of this book, and deal with communication by organizations. What can studying plants teach us about communication? Although the field of communication regularly borrows from other social sciences, relatively few theoretical concepts have been imported from natural sciences. However, Joshua Anderson, Matthew Eastin and Megan LaMonica believe that these fields could offer fertile ground for research into communicational problems. In Chapter 10, they invite the reader to delve into this idea, examining the relationship with uncertainty in environmental strategic communication. The authors draw upon concepts from biology – notably the theory of natural selection, and the r/K selection theory – in order to better understand how organizations can adapt their communication strategies in an uncertain environment. Like organisms which develop survival strategies to deal with unpredictable environments, communicators can spread their communication efforts in order to maximize their impact depending on the degree of uncertainty. In other words, by integrating ecological approaches to communication planning, the authors suggest strategies to adapt to the dynamic and sometimes unpredictable challenges of public engagement with environmental issues. Thus, plants' reproductive strategy can offer a new framework in which to view and understand communication processes.

Finally, Céline Pascual Espuny and Catherine Loneux look at the creation of narratives surrounding biodiversity in discourse on CSR (corporate social responsibility) in organizations. Two "megatopics" are now becoming prominent on corporate agendas: the climate emergency and the biological emergency. While the former has already been an area of concern for the past twenty years, the latter has only become visible more recently, and has not yet been well integrated into corporate strategies. Thus, there are numerous questions over how to grasp and incorporate biodiversity into corporate communication and narratives surrounding CSR. After presenting the socio-communicational context, Chapter 11 offers a critical view of the hybridization of discourse, notably analyzing the emergence of public problems for corporate communication and narrative strategies in the area of CSR. The authors conclude by highlighting the main features of these "narrative megatopics" and the dangers of "biodiversity-washing" associated with them.

The diversity and richness of the contributions presented in this book shed new light on the multiple facets of relations between humans and non-humans – in particular, in the emerging context of communication on plants. By exploring ways in which these interactions can be reframed, the book encourages us to rethink our understanding of the communicational processes that go hand in hand with, and reshape, the discussion of the nature/culture dichotomy, and the traditional boundaries between species. The studies presented here elucidate not only the complex mechanisms by which plants interact with their environment and with one another, but also how such exchanges could enrich our own communication practices. This journey makes us consider the ethical and ecological implications or our interactions with the plant world, broadening our thinking about the contemporary issues of communication in an interconnected ecosystem. Ultimately, this book invites a re-evaluation of the voice of plants in scientific and social discourse, while also calling for a more inclusive and respectful approach to the web of connections which bind the whole of the living world together.

References

Amselle, J.-L. (2013). *Psychotropiques. La fièvre de l'ayahuasca en forêt amazonienne*. Albin Michel, Paris.

Arnoud, E. and Price, L. (1993). River magic: Extraordinary experience and the extended service encounter. *Journal of Consumer Research*, 20(1), 24–45.

Arnould, P., Le Lay, Y.-F., Dodane, C., Méliani, I. (2011). La nature en ville : l'improbable biodiversité. *Géographie, Économie, Société*, 13, 45–68 [Online]. Available at: https://shs.hal.science/halshs-00862457v1.

Barraud, R., Andreu-Boussut, V., Chadenas, C., Portal, C., Guyot, S. (2019). Ensauvagement et ré-ensauvagement de l'Europe : controverse et postures scientifiques. *Bulletin de l'association de géographes français*, 96(2). doi: 10.4000/bagf.5141.

Batke, S.P., Dallimore, T., Bostock, J.I. (2020). Understanding plant blindness – Students' inherent interest of plants in higher education. *Journal of Plant Sciences*, 8(4), 98–105.

Calvo, P. and Keijzer, F.A. (2009). Cognition in plants. In *Plant-Environment Interactions: Behavioral Perspective*, Baluška, F. (ed.). Springer, Berlin, Heidelberg [Online]. Available at: https://research.rug.nl/en/publications/cognition-in-plants-2/.

Calvo, P., Gagliano, M., Souza, G.M., Trewavas, A.-J. (2020). Plants are intelligent, here's how. *Ann. Bot.*, 125(1), 11–28. doi: 10.1093/aob/mcz155.

Catellani, A., Pascual Espuny, C., Malibabo Lavu, P., Jalenques Vigouroux, B. (2019). Les recherches en communication environnementale. *Communication*, 36(2). doi: 10.4000/communication.10559.

Catellani, A., Cholet, C., Pascual Espuny, C. (2023). The intelligence of plants: A new narrative? Notes on a contemporary cultural phenomenon. *Environmental Communication*, 17(6), 618–633. doi: 10.1080/17524032.2023.2234654.

Chamovitz, D. (2017). *What a Plant Knows: A Field Guide to the Senses*. Farrar, Straus and Giroux, New York.

Descola, P. (2005). *Par-delà nature et culture*. Gallimard, Paris.

Gagliano, M. (2012). Green symphonies: A call for studies on acoustic communication in plants. *Behavioral Ecology*, 24, 789–796.

Haraway, D. (2009). *Des singes, des cyborgs et des femmes. La réinvention de la nature*. Éditions Jacqueline Chambon, Arles.

Harrison, R. and Sterling, C. (eds) (2020). *Deterritorializing the Future: Heritage in, of and after the Anthropocene*. Open Humanities Press, London.

Jeanneret, Y. (1994). *Écrire la Science. Formes et enjeux de la vulgarisation*. PUF, Paris.

Kazic, D. (2022). *Quand les plantes n'en font qu'à leur tête. Concevoir un monde sans production ni économie*. La Découverte, Paris.

Kohn, E. (2017). *Comment pensent les forêts : vers une anthropologie au-delà de l'humain*. Zones Sensibles, Brussels.

Lombardi, D. (2021). Se retrouver, c'est déjà se soigner. Processus thérapeutique, ou de la découverte médiatisée de soi, à travers des pratiques néo-chamaniques. PhD Thesis, Université de Paris, 65–82.

Mancuso, S. and Viola, A. (2018). *L'intelligence des plantes*. Albin Michel, Paris.

Morizot, B. (2017). Nouvelles alliances avec la terre. Une cohabitation diplomatique avec le vivant. *Tracés, Revue de Sciences humaines*, 33, 73–96. doi: 10.4000/traces.7001.

Morizot, B. (2020). *Manières d'être vivant. Enquêtes sur la vie à travers nous*. Actes Sud, Arles.

Parise, A.G., Gagliano, M., Souza, G.M. (2020). Extended cognition in plants: Is it possible? *Plant Signaling & Behavior*, 15(2). doi: 10.1080/15592324.2019.1710661.

Peeters, H. and Charlier, P. (1999). Contributions à une théorie du dispositif. *Hermès, La Revue*, 25(3), 15–23.

Pineau, C. (2019). *La corne de vache et le microscope : le vin "nature", entre sciences, croyances et radicalités*. La Découverte, Paris.

Prodi, G. (2021). *The Material Bases of Meaning*. University of Tartu Press, Tartu [Original published 1977].

Sebeok, T.A. (2001). Biosemiotics: Its roots, proliferation, and prospects. *Semiotica*, 134(1/4), 61–78.

Tassin, J. (2016). Les espèces invasives. *Revue juridique de l'environnement*, 41(3), 497–507.

Trewavas, A.J. (2016). Intelligence, cognition, and language of green plants. *Frontiers in Psychology*, 7. doi: 10.3389/fpsyg.2016.00588.

Wandersee, J. and Schussler, E. (2001). Toward a theory of plant blindness. *Plant Science Bulletin, The Botanical Society of America*, 47, 2–8.

Wohlleben, P. (2017). *The Hidden Life of Trees: What They Feel, How They Communicate, Discoveries from A Secret World*. Greystone Books, Vancouver.

Zacklad, M. (2013). Le travail de management en tant qu'activité de cadrage et de recadrage du contexte des transactions coopératives. *Activités*, 10(1), 192–220.

Zhong Mengual, E. and Morizot, B. (2018). L'illisibilité du paysage. Enquête sur la crise écologique comme crise de la sensibilité. *Nouvelle revue d'esthétique*, 22(2), 87–96. doi: 10.3917/nre.022.0087.

PART 1

Semiotic Approaches to the Diversity of Forms of Discourse About and From the Plant World

1

Inclusion of the Plant Kingdom in Communication About Reforesting: Semiotic Analysis

1.1. Introduction

This chapter presents some of the results from a research project on the way in which plant beings – forests, in particular – are present in communications from organizations, in the specific case of organizations devoted to promoting reforesting. Researchers have previously studied communications involving forests and about forests – for example, in relation to promoting reforestation among landowners (Chung et al. 2020) and communications relating to sustainability in the forestry sector (Korhonen et al. 2016). Certain works of research have focused on the way in which forest landowners are presented with information about forest management (Häggqvist et al. 2014); other researchers work on the relationship between human beings and forests (Farcy et al. 2018) and on the shifting treatment of the topic of plant intelligence in media productions (Catellani et al. 2023). There have been few publications on communications to the public about reforestation, aimed at garnering support and financial resources and raising awareness. That is the subject of this opening chapter.

Ecological NGOs and other types of organizations focusing on reforestation develop forms of strategic communication. Here, we will use the definition of strategic communication given by Hallahan et al. (2007):

Chapter written by Andrea CATELLANI.

"purposeful use of communication by an organization to fulfill its mission" (ibid. pp. 3–4). Oliveira et al. (2019) enrich that definition, which is based on a functionalist approach, by integrating other approaches to the analysis of strategic communications in what they call "non-profit communication" (including organizations such as churches, trade unions, foundations, etc.). One of the approaches cited by Oliveira et al. (2019) is linked to an emphasis on "performativity and emerging dimension of the language" (ibid., p. 4). This attention to language is important, but we must also consider the role of other semiotic systems, such as visuals, interacting with the written texts.

Here, we present a multimodal semiotic analysis (verbal and non-verbal visual elements) of the strategic communication developed in support of reforestation projects. Our focus is on the following question: How are plant beings – in particular, the collective entity known as a "forest" – put to use in the multimodal promotional discourse of organizations? In the discourse and images intended to spur recipients to action, to protect and restore the forest, what role does the forest itself play? How are plant beings "enrolled"? This term resonates with the research question addressed by Uggla and Olausson (2013) on the enrollment of nature in tourist information.

We will focus on a French organization which is particularly prominent in the field: Reforest'Action, which has enacted thousands of reforesting projects all over the world. For the analysis, we compiled a diverse corpus comprising web pages, reports and publications on a selection of digital social networks (mainly Facebook and Instagram). In this chapter, we only present the results pertaining to websites and downloadable reports. This analysis is also intended to demonstrate the usefulness of semiotic tools for the intelligibility of strategic communication in studies on environmental communication (Evans Comfort and Eun Park 2018; Catellani et al. 2019; Pezzullo and Cox 2021; Takahashi et al. 2022; Catellani et al. 2023).

1.2. Corpus and methodology details

Reforest'Action is an enterprise (and therefore *is* for profit), certified as a B Corp (a label which attests to "a good level of performance on social, societal and environmental criteria" – see www.reforestaction.com/en/governance), founded in France in 2010, with the stated mission to "regenerate the Earth's ecosystems at a large scale". A large proportion of the body's activities, though not all, can be summarized into two broad areas of activity: firstly, collecting resources from people and enterprises wishing to offset their greenhouse gas emissions; and secondly, in collaboration with 800 local partners, organizing actions to protect vegetation and, in particular, carry out reforestation, in numerous countries, using those resources (1,650 projects in 45 countries since 2012, according to the organization's own website). The site is full of discourse explaining, justifying and celebrating such action, using a particularly interesting set of discursive tools – notably in the context of "carbon offset", which is considered by some to be a smokescreen for failure to take action on the climate[2]. Here, we will simply touch on certain aspects of this wealth of discursive strategy.

The observations presented here represent one step in a potentially much broader project, which could include interviews with members of the organization, analysis of the discourse published on digital social networks and an international comparison, which could include other interesting examples such as that of the search engine Ecosia, which promotes its reforestation activities linked to searches carried out by its users. Thus, this chapter is an initial and exploratory contribution, focusing on a corpus of 18 pages from the Website of Reforest'Action (reforestaction.com), a short-form brochure and a longer pamphlet version (both bearing the same title: "Comprendre, sensibiliser, agir pour les forêts" (Understanding, raising awareness, and acting for forests, 2020)), and an in-depth report ("Notre avenir s'appelle forêt" (Our Future is Called Forest 2018)). The documents were downloaded in 2022.

In order to analyze the presence of plants in Reforest'Action's communications, we apply a toolkit of analytic, semiotic and rhetorical devices, which has been used in previous studies (Catellani 2011, 2018; Catellani et al. 2023). The emphasis is on the narrative aspects, with the application of narrative semiotics as developed by Greimas and Courtés (1979) and Floch (1990, 1995). We also explore the visual aspects of this enrollment of forests (Catellani 2018): How are the iconic and plastic aspects of images used in promoting the organization's actions and its end goals, such as fighting global warming?

1 See: www.france24.com/fr/info-en-continu/20230227-les-cr%C3%A9dits-carbone-un-outil-controvers%C3%A9-pour-lutter-contre-la-d%C3%A9forestation [Accessed January 2, 2024].

We consulted the documents without regard for the technosemiotic peculiarities of the different media, but focusing instead on their semiotic composition (without measuring frequency), to identify part of the existing relationships between the verbal and visual signifiers and the contents or possible signifieds. The main aspect taken into consideration was defined on the basis of the research question, centering on the means of mobilization of plants – notably trees and forests – within the multimodal discourse. The following semiotic aspects were explored:

– Narrative dimension: the "actantial roles" which emerge, and notably those which are assigned to plants. We refer to Greimas' original actantial model (Greimas and Courtés 1979), which involves identifying subjects, objects (of value), senders, receivers, helpers and opponents.

– Axiological dimension: the values and forms of valorization linked to the discursive and visual contents of the pages, notably to see how plants are portrayed in the discourse as entities endowed with values. In particular, we refer to the distinction between the intrinsic values and instrumental values attached to nature, as proposed by Bonnet (2003), with the addition of relational values which are subsequently introduced (Bogert et al. 2022). Relational values refer to the relationship between we humans and the natural world, as opposed to only what we can extract from nature: this is indicative of a more balanced relationship with the non-human, which is no longer viewed merely as a resource to be exploited, but is not yet viewed as holding value in and of itself, in isolation. We also drew inspiration from Greimas' older distinction between modal values and basic values, between what is valued as a tool and what has a value in its own right. The four forms of brand value proposed by Floch (1990) on that basis were also a source of inspiration for the current study: practical or technical, mythological or identity-based, critical or cost/benefit and finally fun or aesthetic, brand value.

– Discursive actorialization of plants: how do plants appear in the discourse? Is it a collection of plants such as a forest, plant species or individual plants, for example?

– The temporal structure of the representation of action in favor of plant life – notably, still images present in the documents or on the web pages. What temporal aspect of the action is represented: the start (inchoative aspect), the action in the process of taking place (durativity) or the effects of the action (final aspect)?

– Finally, we sought to identify overarching forms of "impression" or "illusion" – general rhetorical strategies which seem to operate across all of the media analyzed. Here, we drew inspiration from the concepts of "referential illusion" and "enunciative illusion" proposed by Greimas and Courtès (1979), while not adopting the strictly immanentist view of the text attached to those concepts: "the two references one uses in order to go beyond the microcosm of language, to latch onto an external reality – reference to the subject (to the instance of enunciation) and reference to the object (to the world around us, as an external referent) – ultimately produce only illusion: referential illusion and enunciative illusion".

1.3. Narrative and axiological aspects: the forest as a primary giver and receiver

The pamphlet "Comprendre, sensibiliser, agir pour les forêts" (2020) provides interesting examples relating to the construction of a narrative universe including both humans and plants as actors. Merely by looking at its Table of Contents, we can already begin to identify the actantial structure of that universe. After presenting the "Superpowers of the forest", the text goes on to talk about the "threats" that it faces (narrative complication, to use the elementary narrative framework developed by Larivaille (1974)), before identifying the necessity to take action on a daily basis to preserve the forests and carry out reforestation, and even expand the forests.

Thus, here and elsewhere in the corpus, the forest is presented as the "primary provider and originator", we might say, the source of "services" and endowed with "multifunctionality" on the environmental, economic and social levels ("the cradle of life, the home of health and wellbeing, and the place for leisure", p. 28 of the pamphlet). This "fundamental narrative program" of the forest sometimes brings about an act of recognition: Reforest'Action invites readers to reconnect with this provident giver ("I go to the forest whenever I can for a walk", p. 20), and give it thanks ("In my day-to-day life, I am aware of all the things that surround me and that I use which come from trees: coffee, chocolate, printing paper, toilet paper (yep!), tables, chairs, flooring, and indeed anything made out of wood […] and I give thanks to the forest for all that it gives me"). The forest is also presented as being the victim of deforestation, and the recipient of help from heroes (the heroes in this case being Reforest'Action and the readers of the text, working shoulder to shoulder), as an object with a modal value, so we do not lose the benefits we derive from it.

Thus, Reforest'Action is presented as a heroic figure, standing above all others (though the pamphlet does also make mention of other actors, such as Ecosia), championing reforestation and conservation. However, it is also presented as a receiver and a helper to "us", the readers (individuals and businesses) wishing to take action to protect that source of benefits and services. This positive figure of the enterprise is constructed by weaving a narrative, as well as on a discursive level through the use of figures and data (such as the number of projects carried out, prominently featured on the website on the "Projects" page), and notably through the intensive construction of an ethos of knowhow specific to its teams and advisers. On this point, the pages "Our expertise" and "Ecosystem & Partners" are examples of compilations of skills and expert figures (organizations as well as scientists, listing their qualifications and showing a portrait – a veritable "image-ethos"). Technical jargon (e.g. "endemic species", "carbon sink", "silviculture", "recreating multifunctional forests") is freely used to construct a sort of "ethos of expertise", in addition to the ethos of commitment (see Catellani and Errecart 2021).

Humans and ecosystems are obviously the main beneficiaries of the actions of the enterprise, which clearly sets out the cascade of benefits which come from its reforestation activities. The following is one example among several:

> In the long term, the future forest made of native species will help limit soil erosion, restore the ecological properties of the land, and ensure the resilience of the stand in the face of possible climatic or biological hazards. In terms of biodiversity, the presence of trees, including a fruit species, will contribute to the development of habitats for the fauna and flora. In this rural area, opportunities for the population to increase their standard of living are rare. The nursery work, planting, and maintenance of the trees will provide employment opportunities for local people. Once mature, the introduced species should allow the development of an economy based on the trade of tree resources.

Finally, the individuals and businesses receiving the communication of Reforest'Action are depicted as delegate subjects or helpers of the program in defense of forests and the natural world, called upon to play their part through eco-gestures (which are listed in the pamphlet on meat consumption, leather, chocolate, palm oil, etc.), and by funding reforestation work.

On the pages devoted to ongoing or forthcoming projects, the discourse also paints the picture of the valued figure of local organizations who partner with Reforest'Action. Such is the case of Planalto, described as "the model of a productive and sustainable enterprise" on the page presenting a reforestation project in Mexico (see: www.reforestaction.com/palizada-mexique).

Figure 1.1 shows an example of the discursive exposition of the benefits of forests by visual accumulation, in a didactic fashion, with simplified images (sorts of pictograms) being combined with the figures. On an axiological level, then, the forest is endowed primarily with instrumental and multiple value (the economic, social and environmental aspects), which echoes the three dimensions which have long been recognized as being part of "sustainable development". Indeed, the aim of reforestation is to:

> Restore a multifunctional forest, based on biodiversity in the form of different species of trees, which will provide a wide range of ecosystem services for the wellbeing of all living organisms – humans in particular.

This multifunctionality, then, is also ecosystemic, including living beings, with humans at the center: we could speak of anthropocentric systemic instrumentalization. It is also a discourse of reconciliation of differing interests and values, including those of the timber industry, society, businesses, local communities, landowners and ecosystems. This is a particularly extensive "collective" (in the sense of the term used by Bruno Latour), brought together in harmony in the discourse of Reforest'Action, which thus presents itself as a mediator, a "pivotal point" and a conciliator. This reconciliation of values and interests is found widely in the organization's discourse, notably on the web pages which present current or coming projects, where the environmental, economic and human/societal aspects come together.

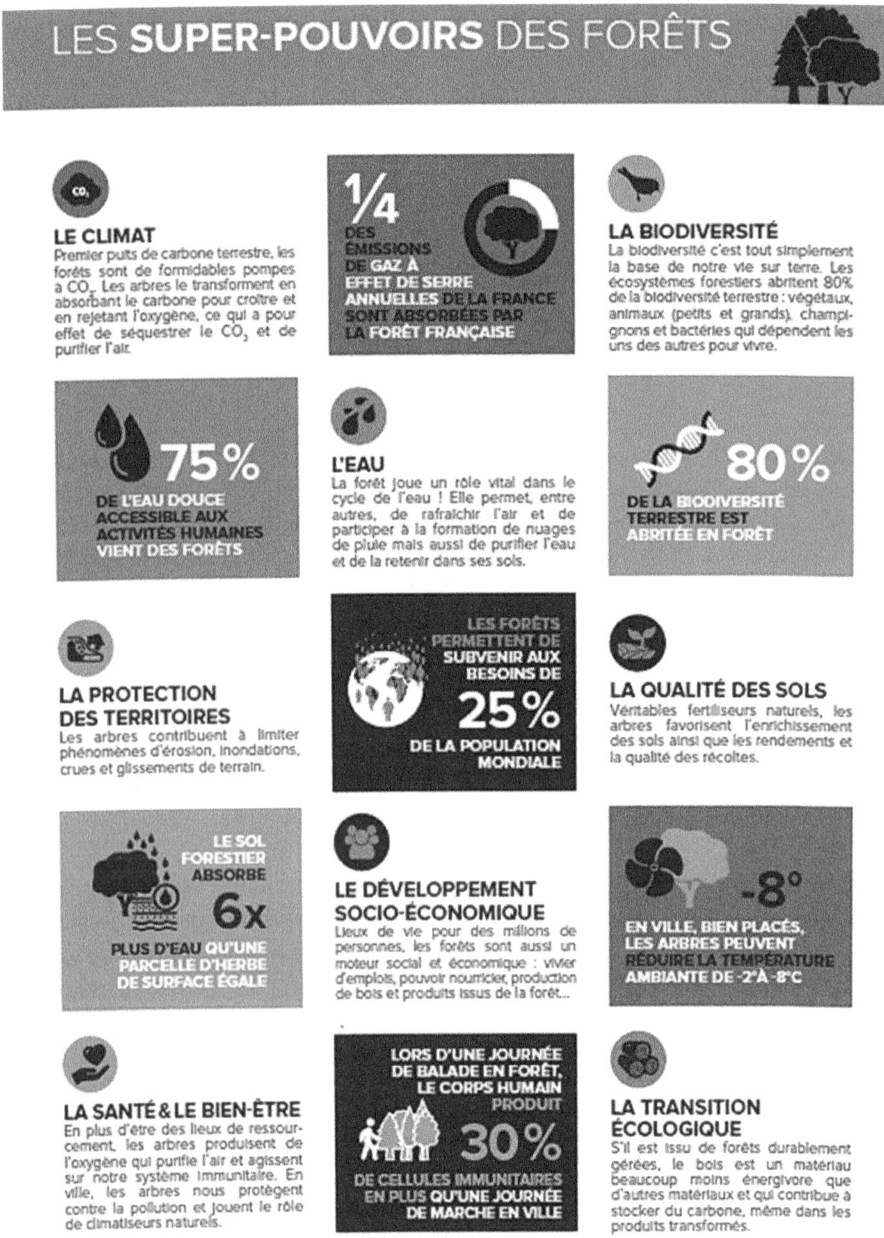

Figure 1.1. *Page taken from the pamphlet "Les superpouvoirs de la forêt" [The superpowers of the forest] (n.d.). Reproduced with the kind permission of Reforest'Action. For a color version of this figure, see: www.iste.co.uk/souillard/emerging.zip*

The domination of an instrumental axiological logic is tempered by the appearance of cases which point to a more relational kind of valorization. When we are invited to "give thanks to the forest" (see above), we can see the emergence of a more balanced, almost spiritual form of relationship. The following testimony comes from a forest landowner who works in partnership with Reforest'Action (see: https://www.reforestaction.com/magazine/parole-de-proprietaire-forestier-jean-louis-bignaud – only available in French), and appears to offer a glimpse of this relational dimension: "From my perspective, the forest is by no means purely economic in function. Every day, I look to it for the pleasure of taking refuge there, admiring its biodiversity, and picking mushrooms." This extract deliberately reveals the aesthetic and relational value which Bignaud attaches to the forest. However, we found no examples of purely intrinsic valorization, although our exploration was not exhaustive. Finally, it should be noted that the trend of compiling and reconciling benefits is very visible in the presentation of ongoing projects or ones soon to be funded (the "Projects" tab), where the impacts are always multiple and multidimensional, and several of the United Nations' Sustainable Development Goals (SDG) are mentioned in relation to each project (see: https://www.reforestaction.com/en/projects/1689). This axiological multifunctionality is added to "reticular" nature on a narrative level (the activation of a network of local actors by Reforest'Action). We could say that the organization is presenting itself as a leading player and an expert in the "city of project" identified by Boltanski and Chiapello (1999), characterized by the centrality of the ability to forge connections.

1.4. Discursive aspects: actorialization and temporalization

The forest and forests, trees in general, or broad categories ("fruiting trees", for example), are the protagonists in the documents consulted for this study. While specific individuals are not examined on a singular basis, or at least not on the pages consulted for this study, sometimes certain species stand out from the crowd as veritable "plant heroes" – the plant world's answer to the animal species which are so emblematic of the communication of associations such as the WWF. Such is the case of the spekboom: a South African tree, which stars on the page presenting a reforestation project (see: https://www.reforestaction.com/en/magazine/tree-species-multiple-benefits-restoration-south-african-ecosystems):

> As a keystone species, the disappearance of the Spekboom in these areas has led to the erosion of a whole range of plant and

> animal species that depend directly or indirectly on it for their development. [...] In the project area, the depletion of the Spekboom is leading to the desertification of the land, the drying up of water resources and the erosion of the biodiversity of the flora and fauna. It is therefore essential to replant this species to obtain the return of biodiversity, the improvement of rainwater infiltration and the restoration of arable land.

The spekboom inhabits the role we saw above, as a primary giver, and it is subject to systemic instrumental valorization. The page gives a detailed presentation of the action of reforestation, focusing on the restoration of this cornerstone of an ecosystem by the subject (Reforest'Action and its local partners). Though the spekboom is enormously important, there are relatively few visual representations of it on the page. The discourse merely shows a number of iconic animal species (elephants, rhinoceri and "exotic" birds), which seem to have a critical place in feeding into the visual imaginary of readers (four out of the seven images on the page show animals). The impression given is that images of plants alone will not suffice: here, animals seem to be enlisted to rescue the plant under threat, to create images that are truly (potentially) impactful. The situation is slightly different on the page in the "magazine" section of the website, which presents the project to restore Albany thickets, again in South Africa[3]. Here, the photographs centering on animals (an elephant, ewes) are in the minority, in comparison to those where plants (notably the spekboom) are the only living species shown, either close up or in the middle distance (six images of plants, in close up or as a panorama, compared to only two with animals at the center, and one showing humans in action in the field, with the plants).

The visual staging of plants – notably the framing of photographs – in their environment takes place in a variety of ways. The first is to show the degraded environment before intervention (deforested land, scorched earth). Next come shots of planting actions, in which the plant is interacting with the humans involved in the project (sometimes children, or groups of villagers) – local actors who play the role of partners, bringing the organization's action to life.

2 See: https://www.reforestaction.com/en/magazine/south-africa-albany-thickets-science [Accessed August 24, 2025].

Figure 1.2. *Image drawn from a reforestation campaign in China (see: https://www.reforestaction.com/en/projects/2788). Reproduced with the kind permission of Reforest'Action. For a color version of this figure, see: www.iste.co.uk/souillard/emerging.zip*

Other times, we may see photos of small trees that have been planted, illustrating a phase immediately following the reforestation action.

We then come to a final type of visual staging which is less present: images which represent the completion phase of the action – the results after a number of years, with a forest growing and thriving after the planting. This type of image is apparently rare: we found only one case, in which a series of photos show "before" and "after" the intervention side by side, for comparison purposes, in relation to one of the organization's older projects. This is an article in the "magazine" section, talking about a project to restore mangroves in Sumatra[4]. This type of image is occasionally used by the search engine Ecosia in the form of narrative videos, to illustrate advances and show the trees which have been planted thanks to searches carried out

3 See: https://www.reforestaction.com/en/magazine/aceh-sumatra-local-communities-restoring-mangrove [Accessed August 24, 2025].

through ecosia.org[5]. The temporal features linked to the narrative of the problem (before) and the remedial action under way, or its early effects (saplings having been planted) are dominant, in a discourse which seems to focus on the past and the present, and often indicate the future only in the form of a promise, or a call to action. The organization explains this trend by pointing to the fact that the reforestation initiatives are fairly recent, and that it takes between 5 and 10 years for the new plants to grow enough to obtain sufficiently "impactful" photos.

Figure 1.3. *Image drawn from a reforestation campaign in Luxembourg (see: https://www.reforestaction.com/en/projects/2948). Reproduced with the kind permission of Reforest'Action. For a color version of this figure, see: www.iste.co.uk/souillard/emerging.zip*

1.5. Overarching rhetorical strategies

The brief observations presented above enable us to identify traces of "overarching rhetorical strategies" in the documents analyzed. This also enables us to compare the case of Reforest'Action to that of two other similar organizations. We can identify an initial type of framing effect: the presence effect, triggered by the proliferation of photos (notably of land needing to be

4 See: https://blog.ecosia.org/fromdrylandtoforestin5years/ [Accessed August 24, 2025].

reforested, nurseries or small plants in the ground) which accompany project reports (photo galleries). These photos, which have little or no verbal "anchorage", to use the term defined by Barthes (1964), nevertheless allow the viewer to feel that they are *there* on the ground; they open up a visual window on a local reality, its vegetation and its human actors who interact with that vegetation. A second type of framing effect, which is more "saturated" with information about the elements shown, is more specific to another French organization devoted to reforestation, *Cœur de forêt* (The Heart of the Forest; see: coeurdeforet.com, only available in French). This organization often adds lengthy captions and texts to photos, giving viewers a deal of precise information. Another variant of the framing effect is shown, in Reforest'Action's communications, in the case of images drawn from geolocation systems such as Google Maps (satellite images of the areas affected by a project). In this case, the framing of the image is more precise, and situates the reforestation action on the scale of the planet, offering an overarching gaze and connoting a type of visual mastery.

The enunciative effect, which is specific to engaging texts which bring us into direct contact with the speaker and their experience, is rare in the communications of Reforest'Action, appearing notably on the pages which provide portrait photos of the members and employees of the association. In many other cases, it is more appropriate to speak of a value-creating staging of the actors involved in the reforestation – notably when they are in the process of planting trees. The communications put out by the Belgian association *Graine de vie* (Seed of Life) use this effect more often, sometimes framing the founder of the association, in the company of local actors, looking us (the viewers) straight in the eye (e.g. see the landing page of https://grainedevie.org/en/about-us/): in this case, the communication becomes more personal, and the organizational ethos is conveyed through the discursive construction of the personal ethos of a charismatic persona.

1.6. Conclusion: making use of the forest in communications

How are plants deployed in the communications of Reforest'Action? On a narrative level, we have identified two main entities: on the one hand, the forest – a primary giver and a provider of "services"; on the other, the enterprise, Reforest'Action, capable of bringing about the beneficial action of a network of actors, the subject of the action, the sender and helper for other actors. We have seen how the forest, as a collective actor, is presented in terms of its modal and (eco)systemic instrumental value. The advantages of the forest are not solely financial: they include benefits for the ecosystem as a whole, and for society. From this point of view, in these texts, we witness the emergence of an extended form of anthropocentrism (Larrère 2016), in a rationalistic framework. While instrumental values seem to dominate, there is an opening up, as we have seen, towards relational values, albeit in a limited way. Thus, if we draw the semiotic square of the types of value attached to the forest, the corpus of texts from Reforest'Action which we analyzed remains focused on the services provided by the forest (forest as a useful tool), with a significant lean towards the economic dimension (forest as a source of profit). On the other hand, there is apparently very limited attention paid to the two poles on the right-hand side of the diagram (forest as a wonder on the aesthetic level, and the forest as a sacred living being).

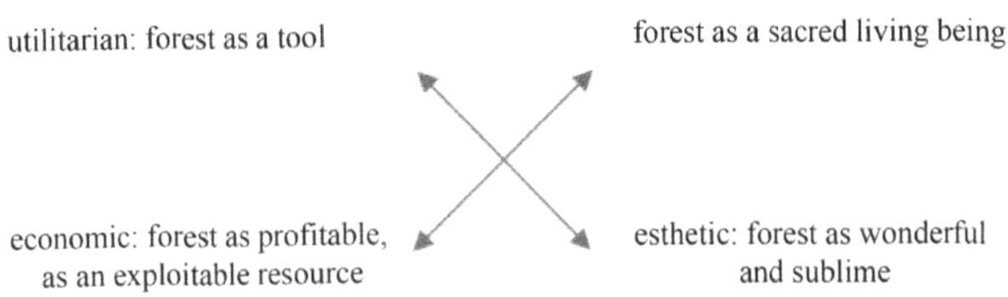

Figure 1.4. *The semiotic square of types of value attached to the forest*

The discourse presents a form of ethos of the organization as an "expert", combined with that of commitment: Reforest'Action gives indicators of being a serious and competent player. It often uses a managerial-style language, taking a stance in the Boltanskian sphere (*cité*, in the Greek sense) of project, and thus positioning itself as a valuable interlocutor for companies wishing to offset their carbon emissions, the ideal mediator to bring the forest into the discourse of management – of capitalism, so to speak – other than in the sphere of individual ecological actions (ecogestures and donations by private individuals). This managerial discursive universe (which we have observed through a qualitative lens, without carrying out lexicometry) makes use of the power of numbers, of accumulation and of efficiency-focused corporate rhetoric.

The forest, and plants, are used in this rhetorical universe – which is at the same time epideictic (praising) and deliberative (aiming to justify action) – as the receiver of an action, valued and understood in rationalized terms. There is a criticism that could be leveled, concerning the domination of the inchoative aspect of the action of reforestation in the images presented. What can be said about the (re)planted forest 3, 5 or 10 years down the line? In the age of the "society of judgement" (D'Almeida 2007), it would undoubtedly be wise to further develop such probative images. However, our analysis was purely an initial, qualitative exploration, inspired by semiotic and rhetorical categories. In addition, Reforest'Action have told us that they are producing in-depth articles tracking the long-term impact of all their projects in Europe[6]. On that page, which is rich in qualitative and quantitative verbal information, photographs taken on the ground show plants and trees that are already quite tall, thus showing a more advanced phase in the interventions.

A fertile field for the study of communicational and organizational practices surrounding reforestation is open to researchers in discourse science and information and communication sciences, to identify the discursive forms in which trees and forests are mobilized in speaking about reforestation.

1.7. References

Barthes, R. (1964). Rhétorique de l'image. *Communication*, 4, 40–51.

Bogert, J.M., Ellers, J., Lewandowsky, S., Balgopal, M.M., Harvey, J.A. (2022). Reviewing the relationship between neoliberal societies and nature: Implications of the industrialized dominant social paradigm for a sustainable future. *Ecology and Society*, 27(2), 7. doi: 10.5751/ES-13134-270207.

Boltanski, L. and Chiapello, E. (1999). *Le nouvel esprit du capitalisme*. Gallimard, Paris.

Bonnet, M. (2003). Nature's intrinsic value. *Journal of Philosophy of Education*, 37(4), 629–639. doi: 10.1111/j.0309-8249.2003.00352.x.

Catellani, A. (2011). Environmentalists NGOs and the construction of the culprit: Semiotic analysis. *Journal of Communication Management*, 15(4), 280–297.

Catellani, A. (2018). Environmentalist multi-modal communication: Semiotic observations on recent campaigns. In *Visual Public Relations: Strategic Communication Beyond Text*, Roberts-Bowman, S. and Collister, S. (eds). Routledge, London, 161–176.

5 See: https://www.reforestaction.com/en/magazine/monitoring-projects-europe-2023 [Accessed August 25, 2025].

Catellani, A. and Errecart, A. (2021). Les “banques éthiques” au prisme des discours sur soi : une identité discursive et narrative hybride. *Communication & langages*, 1(207), 129–147 [Online]. Available at: https://www.reforestaction.com/en/projects/1599.

Catellani, A., Pascual Espuny, C., Malibabo Lavu, P., Jalenques Vigouroux, B. (2019). Les recherches en communication environnementale. *Communication*, 36(2). doi: 10.4000/communication.10559.

Catellani, A., Cholet, C., Pascual Espuny, C. (2023). The intelligence of plants: A new narrative? Notes on a contemporary cultural phenomenon. *Environmental Communication*, 17(6), 618–633. doi: 10.1080/17524032.2023.2234654.

Chung, J.H., Sarmiento, I.G., Van Swol, L.M., Shaw, B.R., Koshollek, A., Ahn, P.H. (2020). Promoting reforestation to landowners: The role of advice-giving through information, efficacy, narratives, and identification in storytelling. *Journal of Forestry*, 118(5), 474–486. doi: 10.1093/jofore/fvaa028.

D’Almeida, N. (2007). *La société du jugement*. Armand Colin, Paris.

Evans Comfort, S. and Eun Park, Y. (2018). On the field of environmental communication: A systematic review of the peer-reviewed literature. *Environmental Communication*, 12(7), 862–875.

Farcy, C., Martinez de Arano, I., Rojas-Briales, E. (2018). *Forestry in the Midst of Global Changes*. CRC Press, Boca Raton.

Floch, J.M. (1990). *Sémiotique, marketing et communication*. PUF, Paris.

Floch, J.M. (1995). *Identités visuelles*. PUF, Paris.

Greimas, A.J. and Courtés, J. (1979). *Sémiotique : dictionnaire raisonné de la théorie du langage*. Hachette, Paris.

Häggqvist, P., Berg Lejon, S., Lidestav, G., Sveriges Lantbruksuniversitet (2014). Look at what they do – a revised approach to communication strategy towards private forest owners. *Scandinavian Journal of Forest Research*, 29(7), 697–706. doi: 10.1080/02827581.2014.960894.

Hallahan, K., Holtzhausen, D., van Ruler, B., Vercic, D., Sriramesh, K. (2007). Defining strategic communication. *International Journal of Strategic Communication*, 1(1), 3–35 [Online]. Available at: https://www.researchgate.net/publication/241730557_Defining_Strategic_Communication.

Korhonen, E., Toppinen, A., Lähtinen, K., Ranacher, L., Werner, A., Stern, T., Kutnar, A. (2016). Communicating forest sector sustainability: Results from four European countries. *Forest Products Journal*, 66(5–6), 362–370. doi: 10.13073/FPJ-D-15-00046.

Larivaille, P. (1974). L'analyse (morpho)logique du récit. *Poétique*, 19, 368–388.

Larrère, C. (2016). L'anthropocentrisme de la bioéthique. *Noesis*, 28. doi: 10.4000/noesis.2953.

Oliveira, E., Melo, A.D., Goncalves, G. (2019). *Strategic Communication for Non-Profit Organisations: Challenges and Alternative Approaches*. Vernon Press, Wilmington, DE [Online]. Available at: https://storage.vernonpress.com/files/web/cc8e4cc4-3c3f-4d36-b84c-a04b25fc729b/1475655362.pdf.

Pezzullo, P. and Cox, R. (2021). *Environmental Communication and the Public Sphere*. Sage, London/New York.

Takahashi, B., Metag, J., Thaker, J., Evans Comfort, S. (eds) (2022). *The Handbook of International Trends in Environmental Communication*. Routledge, London.

Uggla, Y. and Olausson, U. (2013). The enrollment of nature in tourist information: Framing urban nature as "the other". *Environmental Communication*, 7(1), 97–112. doi: 10.1080/17524032.2012.745009.

2

Phytosemiotics: Codes, Signs and the Acts of Transformation

2.1. Introduction

In 1909, Jakob von Uexküll published *Umwelt und Innenwelt der Tiere* (*Environment and Inner World of Animals*), a text that stands as one of the earliest endeavors to comprehend the inner and subjective world of animals, departing from the classical objective external perspective within the methodological framework of empirical science. The pivotal theme of his work is what we could now define as the *agency* of living beings (Latour 1993, 2017; Tønnessen 2015; Sharov and Tønnessen 2021), specifically, the foundation of the capacity to act within a given environment. Of particular significance is Uexküll's research approach, wherein he closely relates the objects of the surrounding world, the inner world of the animal as an agent and the actions of the latter. Thus, two realities that previously seemed separate – the animal and its environment – become a unified structure (Brentari 2015; Padial 2016; Zengiaro 2024). Moreover, his attention was directed not so much towards understanding animal consciousness but rather towards what the animal's behavior reveals about the stimuli encountered in its environment.

Building on the legacy left by Jakob v. Uexküll, biosemiotics opened up the theory of meaning and qualitative investigation to the entire living world. In this chapter, we will see that contemporary arguments investigate code, information and signification not only in the animal world but also in the plant world. The first section will investigate how the form of

---

Chapter written by Nicolas ZENGIARO.

self-organization can be re-read in the light of the theory of meaning and information exchange presented by phytosemiotics. It will examine whether terms such as "read", "interpret" and "know" can be applied to plant life without falling into atropomorphism. Afterwards, an issue that remains open in biosemiotics will be presented, namely whether semiosis is possible without interpretability. In particular, we will look at the arguments of Marcello Barbieri and his theory of code as a minimal structure of semiosis together with a case study on the action potential of *Dionaea muscipula* and *Mimosa pudica*. This will serve to show that the action potential is an interpretation tool based on the codification of the energy exchange between plants and the environment. Finally, a philosophical hypothesis will be presented, based on the metaphysics of Emanuele Coccia, regarding the transformation and manipulation of signs by the plant world to shape the environment. In the latter case, it will be argued that meaning arises along the edges of limitrophy between living beings, which hybridize their different existential manifestations in a given environment. This will also make it possible to give back space in the theory of meaning to the ecosystem, starting with phytosemiotics, for a reassessment of coexistence with other plant life forms.

2.2. Phytosemiotics: self-organization and information exchange

Due to the broadening of the boundaries of the animal world by biosemiotics, Martin Krampen (1981, 1986a, 1986b, 1992) observed that in plant life, communication and interpretation processes occur among different organic systems. This field of study concerning plant life has been termed "phytosemiotics"[1]. The initial focus was on the informational level that an organism exchanges with its surrounding environment (Deely 1982, 1986). This attention to the informational level emphasized the organism's selective attitude towards its environment (Kull 2000). Phytosemiotics presents a controversial issue that has emerged as a challenge to Umberto Eco's assertion that there are certain natural boundaries beyond which we cannot speak of sign processes (Eco 1976), and as evidence of growing attention to the field of biosemiotics. This challenge has been pursued by several researchers (Krampen 1997, 2001; Witzany 2006, 2008; Kull 2009, 2023;

1 The term "phytosemiotics" can be found as early as an article by Ferruccio Rossi-Landi (1975, p. 181), in which he wrote that it might be possible for future research to be phytosemiotics that deals with codes and messages in the plant kingdom.

Witzany and Baluška 2012; Faucher 2014; Gare 2022) showing the potential of biosemiotics in reinterpreting the link between codes, information, meaning and interpretation.

In the field of phytosemiotics, the organism's continuous exchange of energy is integral to its informational processes. This energy dissipation maintains the organism in a state far from equilibrium, characterized by negative entropy (Schrödinger 1944). This state not only supports the emergence of new configurations but also facilitates increasing levels of complexity. Essentially, information drives the self-organization of the system. The challenge in this emerging field of study is to decipher how organisms process this information to sustain and regulate these dynamic states (Barbieri 2016). This is where the philosophical context comes into play, precisely at the point where we can ask: What does it mean to detect, select and understand? Can it be said that every organism has a level of cognition or mind?

Beyond the questions that are still under intense debate today, there is a highly significant issue regarding phytosemiotics. The question concerns the justified or unjustified assignment of the term "semiosis" to organic life in general (Emmeche and Kull 2011; Cobley 2016) and, specifically, to plant life. In other words, do phytosemiotics account for the interpretative structures of general semiotics? If interpreting an object means transforming something merely physical into a sign, deciphering it in reference to the agent's life, then it implies that the agent regulates its behavior (through, for example, a strategy) based on what is interpreted. Modifying our behavior or existence based on information obtained for survival means that there is a process of integrating information and matter-energy carrying the message. In accordance with Peirce's semiotics, an object *X* can be read and interpreted by *Z* as long as the reading of message *Y* is inseparable from the process of matter-energy that carries it, i.e. the use of the physical–chemical substrate. In this sense, "reading" the object as a sign could be identified with metabolizing it (Prodi 2010, 2021). The sign can then be an energy gradient, a specific mass, a protein, a particular pH or a precise temperature.

However, there is always something that we cannot eliminate from the attempt to understand the vital significance of living organisms: the anthropomorphism of terms such as "reading", "interpret" and "know" when applied to the world of nonhuman life. In this type of human understanding of what is non-human, concepts of great interest have emerged, such as that of "self-organization"[2] (Andrade 2006). Organic life as self-organizing, referring to living beings in the way an agent exists autonomously, coupled with the external circumstances of the information exchange process it undergoes for survival, has been rediscovered by authors such as Ilya Prigogine, Humberto Maturana, Francisco Varela and Stuart Kauffman, and more recently from enactivist authors (Thompson 2007; Hutto and Myin 2017; Di Paolo et al. 2018).

The notion of information has been used in biology and biosemiotics to discuss organisms living by hierarchies of control dictated by the same informational constraints. As we mentioned earlier, this extends to all levels: from genes to abstract thought (Pattee 2007). For some biosemiotic authors, such as Howard Pattee, there is something common to all biological, psychological and noetic levels: all these structures are meaning-making, or better, a sign-action[3]. Of course, this does not imply that everything possesses a mind[4] (even though some authors relate the information level of a structure's complexity to the level of consciousness). In reality, this indicates that not all semiotic processes belong to the mental plane, but they may belong, for example, to the realm of self-organization through semiotic processes or through an information process that is received and collected by the organism. Or as Alexei Sharov (2013) argued, a "minimal mind" may exist as an instrument used by agents to shape and classify objects. There is nothing that falls into the classic dualistic opposition of the physical versus the mental; on the contrary, deciphering a sign does not have to account exclusively for a mental experience separated from the material world. Subjective experience is not something exclusively mental (dualism) or emerging solely from cerebral activity (monism). Research on the mind has always assumed that "the mind creates sign relations" (Favareau 2007). However, neuroscientist and bio-anthropologist Terrence Deacon (1997) has proposed a new way of looking at the problem of the origin of language, based on Charles S. Peirce's sign theory, suggesting that instead of studying

2 Recently, it has been argued that some basic steps of intracellular self-organization are accomplished through semiotic processes (Beekman and Jochemsen 2023).

3 "But by 'semiosis' I mean, on the contrary, an action, or influence, which is, or involves, a cooperation of *three* subjects, such as a sign, its object, and its interpretant, this tri-relative influence not being in any way resolvable into actions between pairs" (Peirce 1998 [1907], p. 411).

4 The mind is traditionally regarded as a human faculty responsible for conscious experience and intelligent thought and the intentional directionality of our acts. The components of the mind include perception, memory, reason, logic, world modeling, motivation, emotion and attention.

the human mind as a unique and special case, we should first understand how all organisms have subjective experiences. These experiences are not only created by brains, but are part of a larger set of natural relationships.

According to this perspective, the interaction process is semiotic in nature. In this context, phytosemiotics, the study of signs and processes of signification in plants, offers a unique perspective to explore such foundations. We will examine how this foundation can be found in the concept of "sensification" (Zengiaro 2019), an activity, an agency that infuses meaning into the organic world, i.e. the entire biosphere. In other words, when we speak of interactions between self-organizing agents, we are considering that such interactions are not merely mechanical or random, but carry with them a dimension of meaning. This semiotic approach implies that organisms, through their interactions, create and interpret signs, giving rise to a complex fabric of meanings that pervades the entire biosphere. Through "sensification", the biosphere is not seen as a collection of passive entities, but as a dynamic network of agents that produce and exchange signs. Therefore, by examining "sensification" as a metaphysical and ontological basis, we can better understand how living organisms, including plant organisms, contribute to the construction of a world of meanings. This approach not only enriches our understanding of semiotics in the life sciences, but also opens up new perspectives to explore the complexity and depth of biological interactions on a global level.

2.3. Semiosis as encoding: action potential

In the field of phytosemiotics, as well as in the broader context of biosemiotics, a highly significant branch has emerged for the current debate (Favarcau 2010). This research area explores the possibility for an organism to generate meaning that transcends the issue of the mind-dependent sign as outlined in the classical semiotic discourse of the last century. Is it possible for semiotic activity not to involve cognition or the mind? Similar questions, which until recently were ignored or considered absurd, receive relevance concerning the problems raised by the description of biomolecular processes of transmission of genetic information. In semiotics, the possibility of noncognitive subjects interpreting messages was already theorized by Charles Sanders Peirce, who argued that the concept of interpretive semiosis does not imply any subject with mind and brain, but mind and matter are a continuum that cannot be separated. In this way, Peirce argued that there was no sharp line between mind and matter. Instead, matter was solidified mind and therefore consciousness and matter were also part of the same continuum.

Contemporary research on plant life is advancing towards a dynamic idea that extends the concept of mind as an organic interpreter of signals in the environment (Chamovitz 2012). Although the semiotic mind is a useful concept to consider, as Peirce did by defining the mind as an "external sign" – this definition seeks to link the mind to its semiotic substrate. In other words, the semiotic tradition seeks to demonstrate how signs influence and shape the mind. However, this becomes problematic when it comes to investigating life forms with reduced complexity, no nervous system, no locomotion, etc. We believe, however, that the notion of mind can also be decentralized in understanding how semiosis emerges in the realm of the biological, and this is to detach us from a kind of anthropocentric view of semiosis processes. In fact, in this path, we can follow Sharov's argumentation:

> Mind is not a necessary component of agents. Bacteria are examples of mindless agents that operate via elementary signaling processes such as DNA replication, transcription, translation, and molecular sensing. They do not perceive or classify objects in the outside world as humans do; instead, they detect signals that directly control their actions. Direct control, however, may include multiple steps of signal transfer as well as logical gates (Sharov 2013, p. 348).

As a part of this field of study, we found interesting the article by Marcello Barbieri (2007) titled "Is the cell a semiotic system?". What captures our interest is the attempt to disentangle the semiotic context from that of the mind-dependent interpreter, the mind-subject capable of detecting signs of an ongoing semiotic process. This article builds on the initial definitions given by Thomas Sebeok (2001) according to which "there can be no semiosis without interpretability" and that interpretation is "a necessary and sufficient condition for something to be a semiosis". The question stands on the necessary and sufficient condition of interpretation as a semiotic process. Yet, if biosemiotics is concerned with the sign of the living, by "living" we must mean everything that falls within the concept of *bios*. And Barbieri points out that if what Sebeok states were true, this would apply to every structure of the organic world. However, there is an exception according to the author:

> The exception is the genetic code, a set of conventions that does not depend on interpretation because the rules of the genetic code are virtually the same in all organisms and in all circumstances, and that has been true for almost the entire history of life on Earth. This suggests that interpretation was not necessary for the origin of the genetic code and, more in general, that the origin of semiosis was not the origin of interpretation. But can we separate semiosis from interpretation? Here it will be shown that semiosis can indeed be defined without any reference to interpretation because its qualifying feature is coding and coding requires a codemaker, not an interpreter (Barbieri 2007, p. 2).

Is it possible to separate semiosis from interpretation? To explore this further, we need to consider the relationship between sign and meaning as a "third entity". In the genetic code, in fact, there is no mind-dependent interpreter; the meaning of the sign does not refer to a subject, but to the relationship between the sign itself and the meaning it carries. There is therefore no arbitrary third entity, i.e. a metaphysical subject that intervenes and constitutes the relationship between the two terms. Each sign carries its meaning, and this relationship is unique each time it manifests itself (for some reason, not for someone). There is therefore no fixity or, more precisely, no deterministic relationship filtered through a mind. According to Barbieri, the relation is a third entity with a certain degree of independence. However, referring to general semiotics, we realize that our categorizations take a dualistic or tripartite form in response to the question: What is a semiotic system?

Today, the question of what constitutes a semiotic system is still answered by two classical models. Ferdinand de Saussure proposed a model defining a semiotic system as a duality of "signifier and signified". Charles S. Peirce, however, defined it as a triad of "sign, object and interpretant", emphasizing the importance of interpretation in semiosis. Marcel Florkin, in 1974, likened "signifier and signified" to "genotype and phenotype", suggesting that molecular biosemiotics should follow Saussure's dualistic model, viewing biological systems as defined entirely by genes and proteins. Thomas Sebeok, from the 1960s, supported the idea that language has biological roots and adopted Peirce's triadic model, insisting that semiosis requires an "interpreter" and that interpretability is essential for semiosis. Barbieri proposed a third model, defining the cell as a trinity of "genotype, phenotype and ribotype", with the ribotype acting as a "codemaker" rather than an interpreter. This model views the simplest semiotic system as a trinity of "sign, meaning and code", independent of interpretation.

In summary, to investigate the minimal forms of semiosis in biological agents, we can divide the research paradigm into three models:

1. Saussure–Florkin's duality of genotype and phenotype.
2. Peirce–Sebeok's trinity of genotype, phenotype and interpreter.
3. Barbieri's trinity of genotype, phenotype and codemaker.

These models differ significantly, particularly regarding the necessity of interpretation in defining semiotic systems. In fact, according to Barbieri, there is an internal codemarker to the system itself that does not belong to the realm of interpretation, as it has been understood in the proposed models. The code seems to be the internal medium to connect the sign and the meaning. As the author states, the interpreter can be maintained in semiotics applied to the human world, to culture, society, but to investigate meaning at the level of life, we must disregard this metaphysical mind-dependent subject that acts as an intermediary (and without which the sign does not become part of the semiotic process). The objection often raised is that such semiotic activities involve a stimulus-response context that does not involve interpretation because the system acts automatically. Therefore, the meaning would decline as it would have a deterministic synthesis by the entities in relation.

Marcello Barbieri argues that protein synthesis, governed by the genetic code, occurs in a totally automatic manner, excluding interpretation. However, this automatism does not imply the absence of meaning, as many cultural processes, including language, demonstrate. In the act of speaking, we apply fixed rules; without these, language would not function. Semiosis, the process that assigns meaning to a sign, is a convention created by a coder. Although a sign can have different meanings in different contexts, thus requiring interpretation, the latter consists of selecting between existing rules, not in creating new meanings. Barbieri notes that some codes are context-dependent and may be genetically determined. However, evolution has favored the acquisition of rules through learning, especially in animal behavior. Initially, animals had genetically determined responses, but they had to learn to interpret signals in order to expand their behavioral repertoire. Consequently, learning constructs a behavioral code whose rules depend on the context.

This argument proposed by Barbieri is significant, especially concerning learning to respond to a signal, understood as learning to interpret that signal. This is precisely where our discussion on phytosemiotics comes into play, starting from the response to different signals posed by the relationship that every life maintains with the environment.

Studies on plant life argue that plants not only move through chemical reactions but also react swiftly to contact with other entities. As early as 1760, Arthur Dobbs, Governor of North Carolina, discovered a plant that moved rapidly when an insect landed on it. This plant is the *Dionaea muscipula*, also known as the *Venus flytrap*, already defined by Darwin as one of the most marvelous plants in the world. The carnivorous plant moves much faster than many other plant species, and it has been discovered that it can count. Electrophysiological experiments, which record the electrical signals of the body, reveal that the action potential (the voltage potential that triggers a rapid movement) of plants belongs to the realm of meaning. In comparison with the animal kingdom, the action potential also indicates how the brain encodes information; this means that evolutionarily it is an extended phenomenon between kingdoms. By "action potential", we mean the electrical potential reached at which there is a rapid reversal of the electrical polarity of the nerve cell membrane, occuring following stimulation. The propagation of the action potential along the nerve fiber membrane (i.e. the transfer of electrical charges) gives rise to the nerve impulse. Under resting conditions, cells show a separation of electrical charges at the plasma membrane, consisting of a surplus of positive charges on the outer surface.

In the experiment conducted by Gregory Gage in the 2017 TED Talk titled "Electrical experiment with plants that count and communicate"[5], it is observed how the Venus flytrap reacts differently based on the duration of an object landing on it. The action potential detected by the electromyograph indicates that the plant somehow perceives an intruder inside its cavity. However, if this intruder does not land for more than a few seconds, the plant does not close in an attempt to trap it. This is because reopening the cavity takes from 24 to 48 hours, during which it dissolves its prey with its released chemical elements. And this process requires a considerable amount of energy. Secondly, because this is a secondary maintenance tool for the plant compared to photosynthesis. Thirdly, the behavior of the Venus flytrap usually indicates that the plant only reopens when the fly inside it is dead,

5 See: https://www.ted.com/talks/greg_gage_electrical_experiments_with_plants_that_count_and_communicate visit 05/02/2024.

so, in a sense, it wants to "be sure" of success before closing the trap. If, however, something lands on it for 20 seconds or more, the same action potential is detected as before, but this time, the trap closes. Thus, there is a decoding of information by the plant, seemingly sensitive information, despite it lacking both a brain and a central nervous system. Unlike animals, plants have evolved not to have organs; however, the absence of organs does not imply a lack of functions.

The behavior, not only of carnivorous plants but also, for example, of the *Mimosa pudica*, exhibits an electrical potential that reacts to contact: when touched, the leaves close, and if the stem is touched, it bends downward for a few seconds. In the experiments presented in the TEDx, the cable to detect electricity is placed around the stem of the Mimosa and on the ground. Electromyograph recordings show how the action potential changes upon contact with a different entity that affects it. In other words, when the plant is touched, touch receptors react by sending a voltage to the stem. Similarly, the recording of an action potential occurs when an animal wants to move the muscles of a limb. This highlights how the plant's action potential encodes sensory or relational information to facilitate movement (Karban and Niiho 1995; Goodrich and Tweedie 2002; Reyes et al. 2002).

> It is now known that many non-neural cells are capable of processing information via ion flows generated by ion channels/pumps and propagated by cell: cell junctions [e.g. cardiac tissue; bone cells]. Simulating neural network-like dynamics, stable bioelectrical gradients among non-excitable cells could store information, memories and exhibit many other properties usually reserved for brains. Plants are no exception, especially when we consider that they have, in fact, well-developed pathways for the effective transmission of information via electrical signals (Gagliano et al. 2014, p. 70).

The ability to communicate using electricity is not a feature exclusive to the animal kingdom. The role of electrical potential waves in remote signaling in response to wounding, for example, has been demonstrated in tomatoes (Rodes et al. 1996) and further examined in *Vicia faba* and barley (Zimmermann et al. 2009). The action potential, whether chemical or electrical, seems to have a universal nature in the organic realm. It could be considered the way all information flows within living bodies.

In the aforementioned experiment, the action potential facilitates communication between different plant species. The third part of the experiment involves the following steps: attaching an action potential recorder to the Venus flytrap, connecting it to the Mimosa pudica, touching the Mimosa and waiting for the action potential to pass from one plant to another in the form of electrical information. In other words, the action potential of the Venus flytrap is taken and sent to all stems of the Mimosa. What is observed is that this recreates the behavior of the Mimosa without directly touching it. It moves solely based on the information derived from the other plant through the expressed action potential in response. For this reason, we can detach ourselves from the general semiotic process that necessarily involves an interpreter to give a real meaning to a sign. Between two communicating plants, there is no subject. There is no mind in the classical sense between communication and the exchange of information between two living beings in the plant kingdom. Yet, the indication of a foreign object on one plant is read by another plant in the same way. As a control, if the first plant, which should send the alarm signal, does not detect a potential danger, the second plant receives a different type of information that does not prompt a reaction. It seems there is an interspecific language measured by electrical, chemical, magnetic field gradients, heavy metals, pathogens, etc. (Blande and Glinwood 2016).

The memory studied in plants by other contemporary authors shows that it is possible to "teach" a plant such as the mimosa not to shut down in the absence of a dangerous stimulus. This not only emphasizes the ability of plants to transform sensations into experiences, but also shows that plants can remember and retain information for 40 days. Furthermore, this exchange of information through action potentials shows that plants possess an integrated signaling system at the genetic, molecular, biochemical and cellular levels (Ginsburg and Jablonka 2009).

When we talk about "signaling", we refer to the process of signal transduction that occurs when a molecule carries a signal that activates a specific cellular receptor, either inside or outside the cell. This receptor triggers a biochemical pathway within the cell, creating a response. Every response, in any biological kingdom, participates in a process of homeostasis and integration of information, aimed at maintaining stability in both internal chemical–physical properties and behavioral aspects. A clear example is the plant Mimosa pudica, known for closing its leaves when touched. Recent studies have shown that Mimosas can be "trained" not to close if the stimulus is harmless. In a study by Gagliano et al. (2014), Mimosas exposed repeatedly to harmless stimuli, like a drop of water, eventually stopped closing their leaves. This suggests that the plant "learned" the stimulus was not a threat. This learning involves complex signaling. When touched, specific receptors trigger biochemical reactions causing the leaves to close. With repeated harmless contact, the signaling system adapts, showing memory and learning. This process, similar to that in other organisms, maintains internal stability and appropriate environmental responses. In summary, signaling in plants, as shown by the Mimosa pudica, demonstrates sophisticated systems to perceive, process and respond to stimuli, maintaining homeostasis and integrating information.

2.4. The signs of transformation: limitrophy and permeability

The question of plant life and its meaningful world, as well as the already mentioned notion of sensification, leads us to a highly significant concept in contemporary philosophy, namely that of "limitrophy". We use "limitrophy" to describe this exploration because it concerns what germinates or grows at the boundary, around the boundary and maintains the boundary. It also involves what nourishes the boundary, generates it, raises it and complicates it (Derrida 2008). The information implicit in biological processes is established by the morpho-specific difference, through which the ecosystem as a whole must come to terms in order to maintain a minimum stability. Information, in this case, is context-dependent and involves systems with the ability to perceive and classify elementary differences and complex patterns of differences, i.e. it involves a semiotic process. This semiotic process extends along the differences that emerge in limitrophy, whose porosity threshold allows differences to be maintained, on the one hand, and integrated, on the other.

This exploration is not about eliminating the boundary but about multiplying its forms and making it more complex. By complicating, thickening, delineating, bending and dividing the line, we enhance and diversify the ways in which meaning is created and understood. In other words, the boundary itself becomes richer and more intricate, allowing for a greater complexity of interactions and interpretations. This complexity leads to more nuanced and multifaceted processes of signification and meaning, enriching our understanding of the world.

In the reevaluation of plants by philosophers, there has been a renewed interest in their adherence to the surrounding world. In the writings of

Emanuele Coccia, we can observe the significance of the plant world as a sign of limitrophy, signifying participation in the world in its entirety

> Plants seem absent, as though lost in a long, deaf, chemical dream. They don't have senses, but they are far from being shut in on themselves: no other being adheres to the world that surrounds it more than plants do. They don't have the eyes or ears that may have allowed them to distinguish the forms of the world and to multiply its image through the iridescence of colors and sounds that we give it. They participate in the world in its totality in everything they meet (Coccia 2019, p. 12).

In this limitrophy between the world and plant life, we can perceive the impossibility of properly separating the plant from the world that embraces it, in a fusion and blending that reinterpret existence based on vital hybridization. It is what can be seen as a continuum in making the world around us meaningful, a kind of sensification of the relationships between living things in the environment. Sensification in the plant world is a concept closely related to semiosis, i.e. the process of producing and interpreting signs. This term describes the ability of plants to perceive, interpret and respond to environmental signals through mechanisms reminiscent of semiotic communication. Sensifisation refers to their ability to "sense" their environment through physical and chemical signals and to respond appropriately.

Plant life is, in a certain way, the environment of itself. Indeed, the limitrophy that also unites animal lives transforms into autotrophy in the realm of plants, signifying nutritional autonomy. Plants convert solar energy, as well as raw matter and its chemistry, into meaning for life. Photosynthesis is a transformation of a foreign language, that of solar energy, into a message for living beings. It is the translation work of plants that shapes living matter. Coccia reasonably argues that plants "make world" [*faire monde*]. They make the world because they not only translate heterogeneous matter into a message for life but also become animal organisms in their transformation into nourishment. Their transformation into nourishment for other forms of life suggests the possibility of observing an intricate message encoded within plants. This message transitions from one form of life to another, akin to a horizontal narrative, revealing deeper layers of interconnectedness among living beings.

If we think about it, the first action that a human expresses upon coming into the world is breathing. Breathing is what allows every terrestrial life to exist in the world. This initial act of breathing indicates an overlap of existential planes, where the air is produced as an existential meaning by plant life, to which we owe our origin. Thus, the metamorphosis we undergo by being immersed in this material produced by plants becomes the genesis of an existential meaning, a sensification, that allows an intrinsic transformation of matter. Therefore, our mind, its rationality, the material substrate of the brain, can exist based on a transformation of the message that plants provide to the life cycle

> Plants are the real mediators: they are the first eyes that appeared and opened themselves onto the world, they are the gaze that came to perceive it in all its forms. The world is, above all, everything the plants could make of it (Coccia 2019, p. 26).

Thus, plant life, as a world of meaning and the construction of a primary sense (or a basilar *sense-making* through phytosemiosis), becomes the subject to which humans must also be accountable. Climatologists, philosophers, anthropologists and scientists discuss this: plant life determines the climate and the environment that hosts us, constituting the world in its complexity. Our existence as a species depends on our interaction with plant life, indicating the overlap between the significance of plants and the sign of our destiny (Zengiaro 2022). We are intrinsically connected, and there exists an overlap that unites the planes of every terrestrial being in one large environment, that of plants. The environment, therefore, is the subject of an agency that can forever alter the fate of humans.

In his 1972 work, Bateson asserts that boundaries must encompass relevant pathways, which extend beyond the confines of the unit in question. He argues that this perspective shifts the concept of "survival" from the survival of a physically bounded entity to the survival of "the system of ideas in circuit", essentially referring to semiotic networks. This underscores the challenges in distinguishing between internal communication processes within cells and organisms (endosemiosis) and external communication processes (exosemiosis), as well as the importance of considering the diachronic–synchronic continuity of these systems (Bruni 2008).

The shift from limitrophy to the permeability of living beings marks a transformation in which living entities become permeable in their forms of life and existence within an environment. This is not simply a change in perception, but a profound transition in which the boundaries between living organisms and their environment become less defined. It means a fundamental change in the way we understand the relationship between organisms and their environment.

In this new paradigm, permeability is not unidirectional, where the environment merely accommodates our existence. Instead, it is bidirectional: we are penetrated by the environment through its constant activity of production, and the energy of plant life also flows through us. This energy, originally processed by plants, flows through us, illustrating a continuous exchange in which living beings dynamically interact with their environment. This concept highlights the mutual interdependence and interconnection between humans and the natural world, emphasizing the fluidity and reciprocity in the exchange of energy and resources within ecosystems.

In this sense, limitrophy traces a transition zone between two systems, representing a crucial area for exploring the communicative dynamics of plants and their interaction with the environment. In the classical theory of phytosemiotics, the focus was on the signs that plant life used as communication tools within plant and animal communities. However, if we wish to draw a broader perspective that takes up the achievements of biosemiotics, trying to go beyond a certain anthropomorphism activated by the classical semiotic notions of "reading", "interpretation", "knowledge", we can apprehend the importance of sign interactions in the formation and maintenance of limitrophy itself in plant life forms.

Plants, according to this perspective, not only communicate through chemical, electrical, etc., signals, but along the threshold of limitrophy and existential permeability that makes up their lifeforms. Phytosemiotics, which helps us understand the dynamics of stress, danger, growth and development phenomena, must account for an immersive understanding of the phenomenon of semiosis from an ecological perspective. Our approach therefore seeks to push beyond the mere description of plant signs to consider their dynamic, embodied and transformative impact on the ecological boundary. This perspective not only enriches our understanding of plant interactions but opens new avenues to explore how plants can be active agents in shaping and managing ecosystems. This is the only way to understand the plant subjectivity that an innovative field such as phytosemiotics must be able to express.

However, it is not that simple because:

> If life is tied indissolubly to fluid environments, this is because the relation between a living being and the world can never be reduced to one of opposition (or objectification) or to one of incorporation (which we experience in nourishment). The most primal relation between the living being and the world is that of reciprocal projection: a movement through which the living being commissions the world with what it must make of its own body and whereby the world, on the contrary, entrusts the living being with the realization of a movement that should have been external to it (Coccia 2019, p. 38).

It is an active penetration for both parties involved, which are in no way oppositional but rather two terrains that fertilize each other in a perspective of meaning. This perspective of sense-making is the genesis of an *existential recycling*. It is the body and space, the subject and the environment, mutually immersing themselves and binding in an eternal translation and transformation of existential signs[6]

> If the world is in all its beings, this means that every being is capable of radically transforming the world. Universal mixture embodies the fact that the world is constantly exposed to the transformation brought about by its components. One need not wait for the Anthropocene to encounter this paradox: it was the plants that, millions of years ago, transformed the world by producing the conditions of possibility of animal life. The "phytocene" is the clearest proof that the world is mixture and that every being of the world [*mondain*] is in the world with the same intensity with which the world is in it (Coccia 2019, p. 77).

6 With this, even though it may sound poetic, I refer to the term "existential intertextuality" (see Zengiaro 2023), which draws on the theories of Julia Kristeva for the former term, while the latter is inspired by the existential semiotics of Eero Tarasti. The notion indicates a contagion, a constant contamination, in which heterogeneous elements of a given environment dialogue with each other in a polyphony. It is about relating life and non-life forms in an ecological text, showing that existential intertextuality can be sought in an ecology of relations.

2.5. Conclusion

In conclusion, this chapter offers an in-depth analysis of phytosemiotics, focusing on self-organization and information exchange in plant systems. In section 2.2, we examined how plants self-organize and communicate through semiotic networks, questioning whether terms such as "read", "interpret" and "know" can be applied to plant life without falling into anthropomorphism. Section 2.3 addresses an open question in biosemiotics: whether semiosis is possible without interpretability. In particular, the arguments of Marcello Barbieri and his theory of the code as the minimal structure of semiosis are analyzed, together with two case studies on the action potential of Dionaea muscipula and Mimosa pudica, showing that the action potential can be an interpretative tool based on the coding of the energy exchange between plants and the environment. Section 2.4 presents a philosophical hypothesis based on the metaphysics of Emanuele Coccia, concerning the transformation and manipulation of signs by the plant world to shape the environment. In this context, it has been argued that meaning emerges along the boundaries of limitrophy between living beings, which hybridize their different existential manifestations in a given environment, restoring space for the theory of meaning within the ecosystem, starting with phytosemiotics, for a re-evaluation of coexistence with other forms of plant life.

Integrating these perspectives, the chapter emphasizes the idea that the survival of trophic networks is based on semiotic networks that transcend physical boundaries, emerging from the difference found in the locality of living forms. For "locality", reference is made to Maran's (2014) notion, which refers to the characteristic of semiotic structures to merge with their surroundings in such a way that they cannot be separated from the environment without significantly altering their structure or the information contained in this structure. This concept aligns with Marcello Barbieri's theories on minimal semiosis and Emanuele Coccia's exploration of plant life, emphasizing semiotic processes' fundamental and pervasive nature. The innovation of this work lies in its holistic approach to understanding the interconnectedness of semiotic interactions at different levels of life, from the cellular to the ecological, reinforcing the idea that semiosis is a continuous and integral part of the evolutionary fabric of life and that it manifests itself in the material interaction embedded in the existential relationship of living things starting with plant life.

2.6. References

Andrade, E. (2006). Más allá de la dualidad "genotipo-fenotipo". Complejidad y autorreferencia. *Ludus Vitalis*, XIV(25), 3–23.

Barbieri, M. (2007). Is the cell a semiotic system? In *Introduction to Biosemiotics. The New Biological Synthesis*, Barbieri, M. (ed.). Springer, Dordrecht, 179–208.

Barbieri, M. (2016). What is information?. *Phil. Trans. R. Soc. A.*, 374. doi: 10.1098/rsta.2015.0060.

Beekman, W. and Jochemsen, H. (2023). Self-organization through semiosis. *Biological Theory*, 18, 90–100.

Blande, J.D. and Glinwood, R. (2016). *Deciphering Chemical Language of Plant Communication (Signaling and Communication in Plants)*. Springer, Heidelberg.

Brentari, C. (2015). *Jakob von Uexküll. The Discovery of the Umwelt between Biosemiotics and Theoretical Biology*. Springer, Berlin.

Bruni, L.E. (2008). Gregory Bateson's relevance to current molecular biology. In *A Legacy for Living Systems. Gregory Bateson as Precursor to Biosemiotics*, Hoffmeyer, J. (ed.). Springer, Dordrecht, 93–120.

Chamovitz, D. (2012). *What a Plant Knows: A Field Guide to the Senses*. Straus and Giroux, New York.

Cobley, P. (2016). *Cultural Implications of Biosemiotics*. Springer, Berlin.

Coccia, E. (2019). *The Life of Plants. A Metaphysics of Mixture*. Polity Press, Cambridge.

Deacon, T. (1997). *The Symbolic Species: The Co-evolution of Language and the Brain*. Penguin, London.

Deely, J. (1982). On the notion of phytosemiotics. In *Semiotics 1982*, Deely, J. and Evans, J. (eds). University Press of America, Lanham, 541–554.

Deely, J. (1986). On the notion of phytosemiotics. In *Frontiers in Semiotics*, Deely, J., Williams, B., Kruse, F.E. (eds). Indiana University Press, Bloomington, 96–103.

Derrida, J. (2008). *The Animal That Therefore I Am*. Fordham University Press, New York.

Di Paolo, E.A., Cuffari, E.C., De Jaegher, H. (2018). *Linguistic Bodies: The Continuity between Life and Language*. MIT Press, Cambridge.

Eco, U. (1976). *A Theory of Semiotics*. Indiana University Press, Bloomington.

Emmeche, C. and Kull, K. (eds) (2011). *Towards A Semiotic Biology. Life is the Action of Signs*. Imperial College Press, London.

Faucher, K. (2014). Phytosemiotics revisited: Botanical behavior and sign transduction. *Semiotica*, 202, 673–688.

Favareau, D. (2007). The evolutionary history of biosemiotics. In *Introduction to Biosemiotics. The New Biological Synthesis*, Barbieri, M. (ed.). Springer, Dordrecht, 1–69.

Favareau, D. (ed.) (2010). *Essential Readings in Biosemiotics: Anthology and Commentary*. Springer, Berlin.

Gagliano, M., Renton, M., Depczynski, M., Mancuso, S. (2014). Experience teaches plants to learn faster and forget slower in environments where it matters. *Oecologia*, 175, 63–72.

Gare, A. (2022). Vegetative semiosis. In *Tunne loodust! Knowing Nature in the Languages of Biosemiotics*, Favareau, D. and Velmezova, E. (eds). UNIL, Lausanne, 137–140.

Ginsburg, S. and Jablonka, E. (2009). Epigenetic learning in non-neural organisms. *Journal of Bioscience*, 34(4), 633–646.

Goodrich, J. and Tweedie, S. (2002). Remembering of things past: Chromatin remodeling in plant development. *Annual Review of Cell Development Biology*, 18, 707–746.

Hutto, D.D. and Myin, E. (2017). *Evolving Enactivism. Basic Minds Meet Content*. MIT Press, Cambridge.

Karban, R. and Niiho, C. (1995). Induced resistance and susceptibility to herbivory: Plant memory and altered plant development. *Ecology*, 76(4), 1220–1225.

Krampen, M. (1981). Phytosemiotics. *Semiotica*, 36(3/4), 187–209.

Krampen, M. (1986a). Phytosemiotics. In *Frontiers in Semiotics*, Deely, J., Williams, B., Kruse, F.E. (eds). Indiana University Press, Bloomington, 83–95.

Krampen, M. (1986b). Phytosemiotics. In *Encyclopedic Dictionary of Semiotics*, Sebeok, T.A. (ed.). Mouton de Gruyter, Berlin, 726–730.

Krampen, M. (1992). Phytosemiotics revisited. In *Biosemiotics: The Semiotic Web 1991*, Sebeok, T.A. and Umiker-Sebeok, J. (eds). Mouton de Gruyter, Berlin, 213–219.

Krampen, M. (1997). Phytosemiosis. In *Semiotics: A Handbook on the Sign-Theoretic Foundations of Nature and Culture*, volume 1, Roland, P., Robering, K., Sebeok, T.A. (eds). Walter de Gruyter, Berlin, 507–522.

Krampen, M. (2001). No plant – no breath. *Semiotica*, 134(1/4), 415–421.

Kull, K. (2000). An introduction to phytosemiotics: Semiotic botany and vegetative sign systems. *Sign Systems Studies*, 28, 326–350.

Kull, K. (2009). Vegetative, animal, and cultural semiosis: The semiotic threshold zones. *Cognitive Semiotics*, 4, 8–27.

Kull, K. (2023). Necessary conditions for semiosis: A study of vegetative subjectivity, or phytosemiotics. In *Approaches to Biosemiotics*, Coca, J.R. and Rodriguez Higuera, C. (eds). Ediciones Universidad de Valladolid, Valladolid, 59–75.

Latour, B. (1993). *We Have Never Been Modern*. Harvard University Press, Cambridge.

Latour, B. (2017). *Facing Gaia. Eight Lectures on the New Climate Regime*. Polity Press, Cambridge.

Maran, T. (2014). Place and sign. Locality as a foundational concept for ecosemiotics. In *Re-Imagining Nature: Environmental Humanities and Ecosemiotics*, Siewers, A. (ed.). Bucknell, Lewisburg, 79–89.

Padial, J.J. (2016). Los vivientes como agentes semióticos: tendencias transformativas de la biología en los siglos XX y XXI. *Naturaleza y Libertad*, 6. doi: 10.24310/NATyLIB.2016.v0i6.6296.

Pattee, H. (2007). The necessity of biosemiotics: Matter-symbol complementary. In *Introduction to Biosemiotics. The New Biological Synthesis*, Barbieri, M. (ed.). Springer, Dordrecht, 115–133.

Peirce, C.S. (1931/1935). *Collected Papers of Charles Sanders Peirce, vol. I–VI edited by Hartshorne, C. and Weiss, 1931–1935; vol. VII–VIII Edited by Burks A.W.* Harvard University Press, Cambridge.

Peirce, C.S. (1998). Pragmatism. In *The Essential Peirce: Selected Philosophical Writings*, volume 2, Houser, N. and Kloesel, C. (eds). Indiana University Press, Bloomington, 398–433 [Original published 1907].

Prodi, G. (2010). Signs and codes in immunology. In *Essential Readings in Biosemiotics. Anthology and Commentary*, Favareau, D. (ed.). Springer, Berlin, 323–336 [Original published 1988].

Prodi, G. (2021). *The Material Bases of Meaning*. University of Tartu Press, Tartu [Original published 1977].

Reyes, J.C., Henning, L., Gruissem, W. (2002). Chromatin-remodeling and memory factors. New regulators of plant development. *Plant Physiology*, 130(3), 1090–1101.

Rodes, J.D., Thain, J.F., Wildon, D.C. (1996). The pathway for systemic electrical signal conduction in the wounded tomato plant. *Planta*, 200, 50–57.

Rossi-Landi, F. (1975). Signs about a master of signs. *Semiotica*, 13, 155–197.

Schrödinger, E. (1944). *What is Life? The Physical Aspect of the Living Cell.* Cambridge University Press, Cambridge.

Sebeok, T.A. (2001). Biosemiotics: Its roots, proliferation, and prospects. *Semiotica*, 134, 61–68.

Sharov, A. (2013). Minimal mind. In *Origin of Mind*, Swan, L. (ed.). Springer, Berlin.

Sharov, A. and Tønnessen, M. (2021). *Semiotic Agency. Science beyond Mechanism*. Springer, Berlin.

Thompson, E. (2007). *Mind in Life: Biology, Phenomenology, and the Sciences of Mind*. Harvard University Press, Cambridge.

Tønnessen, M. (2015). The biosemiotic glossary project: Agent, agency. *Biosemiotics*, 8, 125–143.

von Uexküll, J. (1909). *Umwelt und Innenwelt der Tiere*. Springer, Berlin.

Witzany, G. (2006). Plant communication from biosemiotic perspective. *Plant Signaling & Behavior*, 1(4), 169–178.

Witzany, G. (2008). The biosemiotics of plant communication. *The American Journal of Semiotics*, 24(1), 39–56.

Witzany, G. and Baluška, F. (2012). *Biocommunication of Plants*. Springer, Heidelberg.

Zengiaro, N. (2019). Il linguaggio del corpo : il sensibile come nuovo modello di comunicazione. Animal studies. *Rivista italiana di zooantropologia*, 27, 40–47.

Zengiaro, N. (2022). Ecosemiotics of the city. Designing the post-anthropocene. *European Journal of Creative Practices in Cities and Landscapes (CPCL)*, 5(2), 204–228.

Zengiaro, N. (2023). Latour and biosemiotics. *The Hybrid Notion of Life*, 37, 130–143.

Zengiaro, N. (2024). The circulation of meaning: A biosemiotics perspective on functional circle. *Punctum. International Journal of Semiotics*, 9(2), 109–225.

Zimmermann, M.R., Maischak, H., Mithofer, A., Boland, W., Felle, H.H. (2009). System potentials, a novel electrical long-distance apoplastic signal in plants, induced by wounding. *Plant Physiology*, 149, 1593–1600.

PART 2

Coconstructing a New Relationship with the Living World and Environmental Education

3

"Space to Spare": The Paradoxes of Tourist Communication

3.1. Introduction: the paradoxes of tourist communication

Mass tourism has been growing and spreading since the early 20th century, sometimes attracting virulent social critique (Paquot 2014; Christin 2017). While the growth of the tourism industry shows no signs of slowing down, tourists' requirements have become more demanding, varied and variable. Increasingly, tourism services tend to focus on quality and tourists' needs are expressed in relation to culture and the environment, according to economist Zaoual (2007). However, it is still tricky to strike the right balance between the pleasure of discovery, mobility and preservation of wild environments, as was theorized by Holder (1989). In the view of the Secretary General of the CTO[1], visiting wild regions sparks off a vicious cycle, attracting tourists, who require infrastructure, the equipment for which destroys the wild nature of the very sites they are visiting, leading to the abandonment of these sites, which then decay and leave only a degraded and inhospitable environment. Economics and anthropology researchers Minvielle and Minvielle (2014) looked at the iconographic and discursive systems governing the communication of tourism operators in the Sahara, marketed as a place of adventure – a type of tourist discourse which tends to go beyond the paradoxes plaguing traditional tourist communication.

This chapter examines a different context, as well as different operators – specifically, a travel influencer, sponsored by tourism institutions and sports

Chapter written by Erica LIPPERT.

1 Caribbean Tourism Organisation.

brands. Note that there are many travel influencers on Instagram – the social network which runs on images and framing of ourselves, used mainly by 18–34-year-olds.[2] Through Instagram, which accounts for 63% of public engagement with the tourism/travel sector[3], these new advertisers are now taking the place of communications put out by travel agencies or tourist offices.

The questions which guide our investigation here are as follows: What are the discursive and iconographic tools used in the communication broadcast by a travel and tourism influencer? Do the statements and images mask the destructive aspect of tourism? How are nature and natural landscapes represented in this type of communication? We believe these questions are particularly pertinent, given that promoting "natural" beauty spots on Instagram now leads to excessive numbers of visitors, and damages the sites themselves[4].

From this perspective, primarily through the lens of discourse analysis, the goal is also to expand upon the work of specialist Libaert (2010), who highlighted the utopian nature of environmental communications. While there is "nothing new under the sun", on the subject of the narratives of enchantment of specific economic universes (D'Almeida 2012), what changes is the commodification of nature and the moral values that creep into this type of production.

We begin by outlining the theoretical and methodological frameworks of our research, before presenting the results, and finish with a conclusion in the form of an open-ended discussion.

3.2. Analyzing tourist communication of influencers on Instagram: a few basic points

3.2.1. *Conceptual and contextual markers*

The framing of travel stories for marketing purposes on social media has been steadily on the rise in the past decade. On the one hand, this tourist

2 According to Statista. See: https://www.statista.com/statistics/325587/instagram-global-age-group/.
3 According to the 2023 Kolsquare report, "Influencer Marketing in the Travel Sector in 2023".
4 For instance, this article from *Le Monde* speaks of a "curse". See: https://shorturl.at/eDJL6 [Accessed May 30, 2024].

discourse comes from companies or places wishing to attract customers or visitors. Examples include the Instagram account @australia[5], which has 5.7 million followers; the Instagram account @paris[6]; and the Facebook page New York Only, which has 139,000 followers[7]. Alternatively, such discourse may come from influencers specializing in the fields of tourism and travel, such as Rafa Polinesio (seven million followers on Instagram)[8], or the French-language influencers Anil B[9] and Lily Rose[10].

According to Belamich and Ingarao (2021), influencers "are opinion leaders. They engage their community with their passion and expertise in a particular field such as travel, sport or lifestyle, and with the quality of their content". By building trust with their followers better than brands do with consumers, influencers have now become perfect partners.

New marketing techniques have sprung from these practices, such as story-living (Marty 2021) – a technique inspired by storytelling, but in which the influencer is the protagonist. In short, it is "experiential storytelling" (Marty 2021) in which the influencer, in showing and narrating an experience that they are having, promotes a product, place or service. Analyses relating to experiential tourism show that by promoting landscapes of outstanding beauty, communicational magic occurs (Arnould and Price 1993). However, Instagram is the site which, by allowing the highlighting of textual and photographic iconography, notably with filters and textual captions serving as anchorage or reference (Barthes 1964), becomes stimulating for publication of tourist experiences by most anybody (Highfield 2015).

3.2.2. *Theoretical and methodological markers*

In order to analyze the communication of a travel influencer on Instagram, a hybrid theoretical framework was needed, suited to the (sometimes textual) iconization that takes place on that platform. As our aim was to assess the mechanisms of tourist discourse and the resulting iconic and discursive imaginaries of nature, we chose discourse analysis.

5 See: https://www.instagram.com/australia/ [Accessed July 24, 2023].
6 See: https://www.instagram.com/paris/ [Accessed July 24, 2023].
7 See: https://www.facebook.com/newyorkonly/ [Accessed July 24, 2023].
8 See: https://www.instagram.com/ppteamrafa/ [Accessed July 24, 2023].
9 See: https://www.instagram.com/anilbrancaleoni/?hl=fr [Accessed July 24, 2023].
10 See: https://www.instagram.com/lilyrose/ [Accessed July 24, 2023].

This discipline considers texts and images as discourse which, through words and representations, conveys principles, beliefs and ideologies.

In human and social sciences, sociological studies have highlighted the complexity of the condition of tourists, who are trapped between the dream of travel and the economic logics which control their day-to-day lives (Daum and Girard 2018). Studies in discourse analysis look at the creation of discourse establishing places as being worthy of interest (Lepan and Duhamel 2012; Cometta and Stock 2021). Sociolinguistic research describes ecotourist discourse (Dann 1996). In the media, there are ever-growing numbers of newspaper articles looking at the impact of the "Instagrammization" of places of interest[11].

We draw inspiration from a study by the sociolinguist Dann (1996) which, though focused only on the verbal aspects, is relevant for our object of analysis, in that it picks out the identifiable features of ecotourism discourse[12]. That study highlights the following attributes: (i) a contrast between the simplicity of natural and/or wild environments and the complexity of urban environments; (ii) the act of going to places of natural beauty is escapism – it is a matter of getting back to a place which is described as being "real" – from this, we get an entire lexicon relating to adventure, discovery and getaways; (iii) the tourist experience offers visitors the opportunity for "real living"; (iv) the aforementioned features place emphasis on the quest for distance from others (other tourists, or other people in general) – i.e. the search for solitude; (v) this experience corresponds to the idea of a tourist as someone who is active and enjoys movement; (vi) it is a quest for authenticity: by coming into contact with nature (which is exalted) or with others (foreigners), ecotourists are able to tap into the true depths of their reality; (vii) it is a largely masculinized discourse (meaning that in gendered languages such as French, only the masculine forms of nouns tend to be used in tourist discourse, and the feminine forms ignored); (viii) this discourse steers clear of the issues of the economic and environmental repercussions of tourism for the environment and for local

11 See: https://shorturl.at/bhuEI, https://shorturl.at/rBDV2, https://tourismteacher.com/insta-tourism/, etc. [Accessed July 25, 2023].

12 "Sustainable", "green", "fair" tourism, defined as being "Tourism that takes full account of its current and future economic, social and environmental impacts, addressing the needs of visitors, the industry, the environment and host communities," according to the World Tourism Organization.

people; and (ix) this discourse reinforces the imaginaries and the "old myths of paradise and unspoiled nature".

On a visual level, we add the methodology developed by semiotician Wong (2019), affirming that the symbol carries argumentative weight, and becomes the vector of archetypes. However, it is precisely the archetypes (in the sense of "models", "imaginaries", "symbols") of nature which are of interest to us here. Consequently, beyond the strictly figurative attributes of the iconographic content, we will also look at what they convey on a symbolic level, which is part of the connotative domain.

3.2.3. *Jeanne Map: an "institutional" influencer from Quebec*

There are various types of influencers: from megastars, who have millions of followers, to macro-influencers (between 100,000 and 1 million followers), micro-influencers (between 10,000 and 100,000 followers) and nano-influencers (between 1,000 and 10,000 followers)[13]. In this study, we wish to highlight the work of a female micro-influencer, firstly because the discourse of tourism and adventure is largely masculine, and the French word for tourist, of whatever gender, is masculine (*le touriste*) (Dann 1996; Johnston 2001); and secondly because micro-influencers, whom their audience consider to be more authentic and ethical, are more specialized. In addition, one of the selection criteria for the corpus was to focus on a female *speaker* of French, but who is not actually from France, so that the study is not "francocentric".

We selected the Quebec native Jeanne Rondeau-Ducharme (whose Instagram handle is @jeannemap)[14], who had 75,000 followers at the time of the corpus' extraction: Quebec is not a mass tourism destination, the influencer portrays herself as a sustainable traveler, and at first glance, the images in her posts seem to focus on nature, including plants and animals. At first, her profile appeared to correspond perfectly to the genre of narrative tourism and travel, though her biography, in French, mentioned her work with a tourism bureau (@tourismequebec), as well as with brands of outdoor gear (@arcteryxmontreal, @obersonboutique) and local producers (@biereboreale). In actual fact, these mentions reflect financial and marketing interests. This presentation of herself immediately identified the influencer as being part of

13 According to the site E-Marketing. See: https://shorturl.at/ioY13 [Accessed July 25, 2023].

14 See: https://www.instagram.com/jeannemap/?hl=fr [Accessed July 25, 2023].

the promotion of and value creation for Canadian products. Looking briefly at her profile, the style of photographs corroborates the stereotype of postcard pictures, and affirms the active aspect of her travel, with images that show the influencer (or sometimes others) doing sporting activities, which we examine in detail later on.

We selected all of her posts over the course of a year prior to the outbreak of the Covid-19 pandemic (March 10, 2019 to March 10, 2020), to ensure that global event did not interfere with the touristic content, because the pandemic indeed halted activities, and thus altered the content of posts by individuals working for or with the tourism industry. We then compiled the post captions into a textual corpus, and analyzed them using the semantic analysis software Tropes, in that it highlights the referential universes, which are themes extracted from the semantic values of the words used.

We then analyzed the symbolic content, and classified the images on the basis of the signs/content and symbols appearing in the images, drawing upon Chevalier and Gheerbrant's (2021) *Dictionnaire des symboles*.

Before presenting the results, it is necessary to provide some contextual background, to elucidate the spatiotemporal span and the social situation in which the communication takes place. Destinations in Quebec, which is a vast territory[15], are primarily visited by Quebec residents[16]. The population is concentrated around the St. Lawrence River, which leaves large amounts of space for forests (covering half the territory), lakes and watercourses, all making for an "exceptional quality of the natural environment", according to the province's government[17]. Quebec has become attractive and politically "active" in relation to tourism, beginning in the 1920s, with the development of the automobile. It was then that the first tourist guides began being published (Lambert 2015). These guides are the point of convergence between touristic promotion of a place and the expectations of tourists.

Jeanne Rondeau Ducharme can be defined as a travel influencer, because she works to promote several Government-run operators such as Bonjour

15 Measuring 1,667,441 km^2, Quebec is three times the size of Metropolitan France.
16 In 2022, it attracted 22.1 million tourists (including 16.1 million Quebec residents and 3.6 million from elsewhere in Canada), and 34.7-million day-trippers. Data from the official tourism portal. See: https://shorturl.at/iAIL2 [Accessed July 26, 2023].
17 See: https://www.quebec.ca/gouvernement/portrait-quebec/geographie-territoire [Accessed July 26, 2023].

Québec, the SEPAQ[18] and Tourisme Montérégie. She sets herself apart from other Quebec travel influencers because she focuses on tranquil places that are not often visited[19]. Jeanne Map shows us the territory of Quebec through her own lens, in that she produces a view of it which corresponds to an institutional demand, but she manages to add a personal touch.

According to historian Lambert (2015), since it began, Quebec's tourist communication has been using signs that mark Canadian identity, such as churches, log cabins, crossroads, mountains, hills and lakes.

In addition, this communication propagates an ideal of Quebec's climate and its low population density, inviting visitors to enjoy a variety of activities while also benefiting from a certain solitude. Thus, this article will also allow us to assess whether Jeanne Map's Instagram posts adhere to these iconographic and verbal stereotypes of Quebec's tourist discourse.

3.3. Romanticized storyliving about nature and the "postcardization" of landscapes

3.3.1. *Archetypal visuals of ecotourist communication*

Over the period studied, Jeanne Map's Instagram profile included 91 posts, containing 116 photographs, five iconized texts[20] and one video. The promotion of photographs as a preferred format accentuates the aesthetic of a postcard – that so definitive of tourist objects.

While the wild landscapes are omnipresent, the framing of the influencer herself is equally important. It is storyliving, in that Jeanne Map shows herself in the process of living, experiencing and contemplating the "wild" landscapes which she is promoting. Table 3.1 gives an overview of the content of the images.

18 Société des établissements de plein air au Québec (Society of Quebec Open Air Establishments).

19 According to the magazine, Clark. See: https://www.clarkinfluence.com/post/voyagequebec [Accessed July 26, 2023].

20 According to Paveau (2019), in an iconized text, "textual and iconic codes and coconstructed as a composite entity".

Denotative content of the images	Number of occurrences
Portrait of Jeanne Map	44
Wild landscape with no humans present	19
Architecture within a landscape	16
Urban environment	9

Table 3.1. *Denotative content of the images posted on Instagram by Jeanne Map*

In her storyliving, Jeanne Map mostly uses romanticized images of wild or rural landscapes (free of other human beings) as key visual elements in her shots, even when she herself features in the images. The influencer subjectively describes her experience of the places in which she practices healthy activities. Thus, apart from the places she showcases, she depicts herself as the model of an “ascetic” lifestyle suffused with contemplation, solitude, outdoor sports and relaxation. It is a romantic and Westernized way of conveying landscapes and nature: healthy and aesthetically beautiful, firmly centered in human subjectivity. The images highlight the dichotomy between nature and culture: human beings have come as observers or users of an independent space, which appears untouched, undeveloped, pure. We also find a Rousseauian and romantic conception, firstly because Rousseau advocated walks in the “natural” environment as a means of escape, allowing for cathartic reverie and solitude, and secondly because visiting nature is a way of accessing its truest essence, happiness and simplicity.

The majority of the images are a testament to broadening horizons, to voyaging (both within Quebec and in France), and to the influencer’s testing herself and pushing her boundaries, which she encourages her audience to do too. There are two main trends that can be seen from the visual analysis. Firstly, we see many of the features highlighted by Dann (1996) on the subject of ecotourist discourse, adapted to the reality of Quebec, and of Canada. (1) Spending time in natural and/or wild places enables people to escape from another reality (the other extreme, implicitly, being the urban environment) to rediscover themselves in solitude: Jeanne Map is mostly shown alone within vast landscapes, with unobstructed views. These wild environments and the cycle of the seasons (which she accepts, and enjoys in their own right), highlighting the character of Canada, represent purity (snow), power (peaks, water, large expanses), cold and transcendence (mountains), the infinite (oceans and seas), life (trees), calm (the absence of crowds) and vacations (sun, beaches), and the reddish-russet hues of fall (the Canadian flag, and the maple). (2) The experience suggested seems to be a peaceful and pacifying one: calm is connoted by identical framing in all of the photos, in which the horizon is in a central position, the color scheme moves with the seasons and the whole of her profile is filtered for a homogeneous effect. (3) If she is not shown in contemplative poses, Jeanne Map paints herself as an active and mobile person (enjoying outdoor sports such as fatbiking, snowshoeing, hiking, etc.). The iconography of Quebec and France (the two areas which appeared on the profile over the course of the period under study) is to be expected. The images are shot through with all of the sublime aspects of stereotypes and collective imaginaries surrounding these places: snow, "Indian summer"[21], endless pine forests, skiing, cabin folklore for Quebec; cafe terraces and their identifiable chairs, Hausmann architecture for Paris, permanently snow-capped peaks for the Alps and jagged cliffs for Britanny.

Secondly, these images romanticize Quebec by hiding everything which makes it appear less than sublime, which is reminiscent of the smooth Disneylandization of the world (Brunel 2012). The deliberate framing of nature, for commercial purposes, is based on the influencer's personal relationship with nature, which provides dreams and enchantment. Jeanne Map depicts Quebec's wild landscapes as sacred, almost untouched by human civilization, where the weather and climate are celebrated, in that they cannot be controlled. Thus, the images correspond to the fantasy of the North American wilderness (Strigler 2023). They speak of practices contrary to mass tourism, by completely excluding the traces of industry, machines, technologies, dams[22] – ultimately, everything which comes from industry; that which is dirty, noisy and polluting in human civilization. The means of transport which Jeanne took to reach these places are never shown. The framing and narration eliminate the tools needed to reach her storyliving.

It is a romanticization and idealization of Quebec, corresponding to nostalgic sociodiscursive imaginaries. On this subject, historians Mathieu and

---

21 Today, this expression raises some controversy among Quebec's residents, because it refers to the Autochthonous people, and the fact that climate change has brought about a shift in the cycles, as mentioned by the newspaper Le Devoir. See: https://shorturl.at/pwIL0 [Accessed August 1, 2023].

22 Indeed, it is possible to visit dams for tourism, as promoted, for example, by the Crown corporation Hydro-Québec. See: https://shorturl.at/jmJY6 [Accessed August 1, 2023].

Lacoursière (1991, p. 38) point out that "in Quebec's collective memory, nature [...] has long given people to dream, usually awakening feelings of strength and nostalgia which were expressed in the following terms: immensity, wild forest, unsullied nature, crystal waters, inexhaustible resources, winters of yesteryear". However, in the early 20th century, this narrative was reversed, with what the historian Stéphane Savard (Savard 2013, p. 80) describes as "pride in dominating nature". Thus, in her way, Jeanne Map is reestablishing a memory and a narrative which preceded Quebec's industrialization. Behind the visual dimension of this communication favoring the framing of "unadulterated" nature, we see the principle of consumption of a place for purely health and (pseudo-?)spiritual purposes. The omnipresence of landscapes, plants and animals allows those who travel there to witness authenticity, and to have a privileged experience of life in nature. Nature is represented as being a place to host human pleasures, as well as a communication tool.

3.3.2. *Experience and romanticization of nature*

The textual corpus compiled from the captions of the 91 posts amounts to 3,924 words, which is an average of 43 words per post. Remember that Instagram gives priority to images, to the extent that, from the third line onwards (125 characters), the text is truncated, forcing users to click into it in order to read any further. This tells us not only that Jeanne Map attaches more importance to images, but that her style of communication is perfectly suited to the particular platform. Her communications leave it up to the natural environment she depicts to convey beauty, rather than doing so through words. The anthropologist Geneviève Brisson (Brisson 2006) points out that the figure of the guide, linked to the figure of the Quebec lumberjack, is recognized as having a "near mythological value. Thus, he becomes a mediator, enabling citizens in turn to access that privileged natural state". This is exactly the stance that Jeanne Map adopts, as we will now examine through textual analysis.

On an enunciative level, it is logical that the pronoun *je* [I] (88 occurrences, 45.8% of subject pronouns – for example, "je le visite" [I'm visiting it]; "je poursuis mon périple" [I'm continuing with my trek]; "je redécouvre aujourd'hui" [Today, I'm rediscovering]) is the most widely used. The same is true of subjective adjectives (60%, 136 occurrences – for example, "paradisiaque" [heavenly], "magnifique" [magnificent]), because as an influencer, Jeanne Map's narration hinges on her own experience. The guide conveys her view of the landscapes and the natural environment, and as it is a promotional type of discourse, alongside the placenames ("Gaspésie", "Québec", "Mont Olivine", etc.) and spatial markers ("au loin" [in the distance], "à quelques pas" [a stone's throw away], etc.), we find meliorative vocabulary ("grand" [great], "séduis" [seduces], "belle" [beautiful], "surréel" [surreal], "magique" [magic], "grande beauté" [amazing beauty], etc.), markers of subjectivity and enthusiasm (145 exclamation marks, 20 occurrences of emotion-loaded terms such as "coups de cœur" [smitten], "désir" [desire], "bonheur" [happiness], "fun"), amplifiers in the form of adverbs ("extra-mega-hâte" [I-just-can't-wait]) or emojis ("mon Québec ♡" [my Quebec ♡]).

The communication promotes a multisensory experience of nature. Thus, in the lexical field of cognition, the word "expérience" [experience] is used nine times, and the word "souvenir" [memory] four times. Analyzing the verbs used (ten occurrences of the verb "voir" [see], eight of "découvrir" [discover], six of "suivre" [follow, or pursue], five of "partir" [set off], five of "sentir" [feel]), we obtain an image of Jeanne Map as an active visitor or traveler, energetic and watchful: it represents the opposite of a passive tourist. Thus, her relationship with nature is not restive; it is proactive. Similarly, intensity modalizers accentuate the euphoric aspect of her discourse, the marvel that she experiences, and the numerous emotions she feels, such as impatience, joy or motivation ("surtout hâte" [cannot wait], "j'ai eu la chance d'essayer BEAUCOUP d'activités" [I was lucky enough to try LOADS of activities]).

Visiting wild places is a virtuous thing to do. This ecotourist communication promotes moral values. Visiting wild or "natural" places opens the door to "#sagesse" [wisdom] and authenticity, to understanding of the past ("maisons d'époque" [old-fashioned houses], "un endroit rempli d'histoire" [a place filled with history]), to freedom ("sky's the limit", "vagabonder" [wandering free]), to sobriety ("#voyagerléger" [traveling light]) and good times with friends (she also encourages her audience to "rencontre" [get together]: "rassembler ta gang" [getting the band back together], and "cohabitation"). Jeanne Map also promotes ecotourism by advertising "biodegradable" products (which can be "recycled") or by talking about choices that she makes ("modifier mesh habitudes de consommation" [changing my consumption habits]). Storyliving favors a representation of authentic understanding of nature, of an art of living and falling back in love with the world. The guide puts her own stamp on her relationship with wild environments by adding the touristic dimension and the adventuring aspect of imaginative and fun perspectives: for her, going into nature is a matter of reclaiming the presumed innocence of childhood, of preindustrial times and reconnecting with her inner self. For example, the following statement – using words, hashtags or emojis, emphasize the aesthetic and metaphorical aspects of walking in the forest ("Couleurs d'automne […] 🍂" [Fall colors], "marcher dans les nuages" [walking in the air]), her enthusiasm ("bin oui" [Yeah!], "qui est biiiiin excitée" [is reeeeeeally excited], "c'est sûuuuuuur" [absoluuuuuuuuutely!]), childlike joy ("se jeter dans la barbe à papa gaspésienne 🍭" [diving into Gaspé cotton candy], "retomber en enfance" [back to my childhood]; "forêt magique" [enchanted forest]) and authenticity ("#madeinquebec", "#quebecoriginal").

Finally, note the crucial role played by indexing, tagging and geolocation. Through the sustained use of hashtags (416 occurrences) and arobases [@ tags] (92 occurrences), the influencer simultaneously tends toward transparent communication (she openly acknowledges her commercial partners and the revenue that she receives from them) and product placement, be it a nature-product or a place-product (the geographic location is, of course, mentioned in all her posts).

The commoditization of nature and of wild spaces is primarily achieved through these functions, while the images serve the function of enchantment and attracting. Thus, we find 212 hashtags relating to place (e.g. "#gaspesie" [the Gaspé Peninsula], "#chamonix", "#nature"), 85 tourist slogans (e.g. "#visitbretagne", "#explorecanada"), 77 mentioning the partner company (e.g. "#sepaq"[23], "bellalttv", "#tremblantpartenaire"), 28 expressing a positive emotion or value ("#gaspesiejetaime" [I love the Gaspé Peninsula], "feelrebornhere", "#sagesse" [wisdom]) and finally 14 acknowledging her role in advertising a place or product ("#pub", "#collaboration"). Except for nine mentions of people who are with her, the identification tags refer to partner companies for which she is providing services, or the equipment she is using (e.g. "@tourismegaspesie", "@heli_tremblant", "@coffeedrup", "@jardinsdemetis"). The textual geolocation indicators, finally, serve as a relay to the image, while allowing audience members to click on the hyperlink created by that function and see where in the world the influencer is, on a world map, letting their imaginations do the rest.

23 Société des établissements de plein air du Québec.

3.4. Conclusion: paradoxes of (eco)tourism communication

The sociologist Yaara Benger-Alaluf shows that tourist discourse presents nature, first and foremost, as a place of health and wellbeing. She observes that, since tourism began, "the notions of authenticity, harmony and sensitivity have been attached to nature, while emphasis has been placed more on individual experience, pleasure, feelings and emotion" (Benger-Alaluf 2019, p. 56). Thus, the foregoing analyses demonstrate that Jeanne Map's communication corresponds fully to this anthropocentric way of working, to the point that she has become a stereotypical exemplar of it.

Indeed, we have been able to observe, firstly through visual analysis, that the images she posts are reminiscent of photo postcards – one of the tourist objects by which individuals are able to communicate their impressions of their travels to other people. We saw that the choice of photographs as a preferred medium, the framing, color scheme and denotative content mark Jeanne Map as a travel influencer, whose goal is to promote (herself) (within) destinations. Through her images, she provides an interpretation and a brand of (primarily) the Quebec territory, presenting it in such a way as to extol its wild attributes. The visuals correspond to the imaginaries and symbolic representations of Old-Time Quebec, romanticizing the relationship with nature. The relationship which she establishes with the territory through her communication also includes the promotion of outdoor and sporting activities: Jeanne Map introduces places that are unknown to the general public (aesthetic aspect) and the activities that you can enjoy there (health aspect); she demonstrates topographical and geographical knowledge (picturesque, ecological aspect, to use the expression coined by geographer Donadieu (2007)), and sheds light on alternative tourist practices (faraway places, difficult to access, recyclable gear and a privileged relationship with).

The discourse analysis showed that the uniqueness of Jeanne Map's content lies in a fun and serene relationship with the environment, and with activities that can be undertaken far from cities. All of the discursive attributes of ecotourism discourse highlighted by Dann (1996) were found in Jeanne Map's texts and images. In addition, hashtags, identification tags and geolocation tags embellish and mark the text with positive emotions, authenticity and an air of magic, while conveying the idea that nature can be consumed in the same way as the associated products (gear or locations). Wild spaces are displayed in contrast to the urban environment, given that the photos contain no sign at all of urban civilization. The few manmade objects that appear are tools for sports practices, or resources (skis, spa, isolated log cabins, etc.).

There are three points which should be touched upon in conclusion. Firstly, as Dann (1996) observed, there is no reference at all – either visually or discursively – to the economic and environmental impact of the tourism that Jeanne Map promotes. The guide does not explain how she travels to these faraway places, how tourists can be accommodated there and what infrastructures are available (other than to promote a service such as a spa, a sporting activity or a gite), whether her presence and that of other potential tourists is welcome to the local human and animal population, what impact it would have, etc. Although she says, on the subject of the natural sites she visits, that there is "d'la place en masse pour tout le monde" [space to spare for one and all], by encouraging her audience to occupy these spaces in Quebec and enjoy their high capacity to accommodate visitors, is she not promoting the idea of excess visitor numbers, which is so undeniably specific to mass tourism? In our view, that narrative leads to ultra-idealization of nature and travel, and to the paradoxes mentioned in the introduction.

Secondly, the way in which the images are framed reinforces the imaginaries and the "old myths of paradise and unspoiled nature" (Dann 1996), given that what is primarily put forward are portrait photos of Jeanne Map herself, living in landscapes with very little or no visible development, or photos of landscapes in which there is no architecture, or whatever architecture there is blends into the landscape harmoniously. However, numerous works of research now show that "untouched", "pristine" nature, set apart from human beings, is a mental image formed by a western social construct (Larrère 1997; Descola 2005; Chansigaud 2013), etc., which, as we know, brings problems along with any benefits. Thus, paradoxes arise throughout her discourse because, while Jeanne Map and other influencers of that ilk extol the calmness, authenticity and naturalness of these wild places, in order to maintain those qualities, surely the tourist discourse about the sites should also show the flip side of the coin: the fragility of these environments, the way in which they coexist with industrial infrastructures, the impact that tourism has on local residents, etc. Ultimately, the question is whether the communication emanating from tourism/travel influencers is inevitably doomed to this type of referential paradox, in which nature and culture are two opposing entities.

Finally, the study of the hashtags, geolocation and identification tags ought to be extended, though it has not been here for lack of space. Indeed, these represent the mechanism by which the places are indexed, becoming hyperlinks and being included in a topic-specific discursive community: they are labeled in this manner, and in being labeled, become products. Remember that in this context, the use of identification tags is to mention partner brands or services. In the posts, these brands and services interact with the topic-specific tagging of the spaces, in a dynamic of association [site > product > tourist office/brand]. Thus, the relationship with the wild environment tends to be presented as one of consumption, for commercial purposes. Finally, the hashtags used carry ecological values such as mentioning the name of the local territory (and tagging it), spirituality in harmony with nature ("#sagesse" [wisdom], "#namaste"), sobriety ("#voyagerléger" [traveling light]), but there is dissonance with the universe of associated brands which do not always have good ecological practices ("#Airfrance", "#nespresso"). The above raises questions about the paradoxes and dissonances at play in tourist discourse, whether or not it claims to be for eco-tourism.

3.5. References

Arnoud, E. and Price, L. (1993). River magic: Extraordinary experience and the extended service encounter. *Journal of Consumer Research*, 20(1), 24–45.

Barthes, R. (1964). Rhétorique de l'image. *Communications*, 4, 40–51.

Belamich, S. and Ingarao, A. (eds) (2021). Fiche 26. Évoluer vers une stratégie digitale d'influence. In *Fiches de marketing*. Ellipses, Paris, 225–232.

Benger Alaluf, I.E.Y. (2009). Emotions in consumer studies. In *The Oxford Handbook of Consumption*, Wherry, F.F. and Woodward, I. (eds). Oxford Handbooks, Oxford. doi: 10.1093/oxfordhb/9780190695583.013.10.

Brisson, G. (2006). L'homme des bois d'Anticosti. La figure du guide de chasse et les conceptions sociales de la forêt québécoise. *Revue d'histoire de l'Amérique française*, 60(1–2), 163–189.

Brunel, S. (2012). *Le développement durable*. Presses Universitaires de France, Paris. doi: 10.3917/puf.brune.2012.01.

Chansigaud, V. (2013). *L'homme et la nature. Une histoire mouvementée*. Delachaux et Niestlé, Paris.

Chevalier, J. and Gheerbrant, A. (2021). *Dictionnaire des symboles*, 3rd edition. Bouquins Éditions/Jupiter, Paris.

Christin, R. (2017). *Manuel de l'antitourisme*. Écosociété, Montreal.

Cometta, M. and Stock, M. (2021). Discursive construction of a destination. Urban transition through tourism in Ticino between 1980s and 2010s. *Mondes du Tourisme*, 19. doi: 10.4000/tourisme.3620.

D'Almeida, N. (2012). *Les promesses de la communication. La force de la parole dans la communication d'entreprise*. PUF, Paris.

Dann, G.M.S. (1996). *The Language of Tourism. A Sociolinguistic Perspective*. Cab International, Wallingford.

Daum, T. and Girard, E. (2018). *Du voyage rêvé au tourisme de masse*. CNRS Éditions, Paris.

Descola, P. (2005). *Par-delà nature et culture*. Gallimard, Paris.

Donadieu, P. (2007). Du pittoresque au durable. Les natures touristiques du paysage. *Espaces*, 254, 16–23.

Highfield, T. (2015). Depicting social television on Instagram: Visual social media, participation, and audience experiences of #sbseurovision. Article, Slideshare.

Holder, J.S. (1989). Tourism and the future of Caribbean handicraft. *Tourism Management*, 10(4), 310–314.

Johnston, L. (2001). (Other) bodies and tourism studies. *Annals of Tourism Research*, 28, 180–201.

Kolsquare (2023). Influencer marketing in the travel sector in 2023, hotels, transport, museums, key figures, the major players, the share of voice, inspiring campaigns and trends for 2023 [Online]. Available at: https://www.kolsquare.com/en/free-ressources/influencer-marketing-travel-sector-report#target-form.

Lambert, M.-E. (2015). À travers le pare-brise : la création des territoires touristiques au Québec et en Ontario (1920-1945). *Revue d'histoire de l'Amérique française*, 68(3–4), 271–299.

Larrère, C. (1997). La nature est-elle vierge ou sauvage ? In *Les philosophies de l'environnement*, Larrère, C. (ed.). PUF, Paris, 85–104.

Lepan, L. and Duhamel, P. (2012). Un discours mis en image : Paris à travers les Guides Joanne – Guides bleus (1863 à 2010). Une approche exploratoire et diachronique de l'espace touristique. *Mondes du Tourisme*, 6, 6–22.

Libaert, T. (2010). *Communication et environnement, le pacte impossible.* PUF, Paris.

Marty, S. (2021). "Swipe up" et "codes promo" : quand les influenceurs donnent vie à un storyliving dédié aux marques. *Communication & Management*, 18, 47–65.

Mathieu, J. and Lacoursière, J. (1991). *Les mémoires québécoises.* Presses de l'Université de Laval, Quebec.

Minvielle, J.-P. and Minvielle, N. (2014). Le tourisme expérientiel au Sahara : une construction hyperréelle de l'aventure, Mondes du Tourisme [Online]. Available at: http://journals.openedition.org/tourisme/375 [Accessed 1 December 2023].

Paquot, T. (2014). *Le voyage contre le tourisme.* Eterotopia, Paris.

Paveau, M.-A. (2019). Technographismes en ligne. Énonciation matérielle visuelle et iconisation du texte. *Corela*, HS-28. doi: 10.4000/corela.9185.

Savard, S. (2013). *Hydro-Québec et l'État québécois, 1944–2005.* Éditions du Septentrion, Quebec.

Strigler, M. (2013). La wilderness : un espace fantasmé. In *L'espace du Nouveau Monde. Mythologies et ancrages territoriaux*, Laffont, G., Gautier, A., Martouzet, D., Chamerois, G., Bernard, N. (eds). Presses Universitaires de Rennes, Rennes.

Wong, M. (2019). *Multimodal Communication. A Social Semiotic Approach to Text and Image in Print and Digital Media.* Palgrave McMillan, London.

Zaoual, H. (2007). Du tourisme de masse au tourisme situe : quelles transitions ? *Marché et organisations*, 3, 155–182.

4

Human/Plant Connections in Mid-mountain Regions: From Resource to Relatedness?

4.1. Plants and the public sphere

Though we have seen vegetation shrinking in urban areas (admittedly, less so since the Covid-19 crisis), over the past several decades, plants have been regaining their place in the perception of our relationship with the world, featuring in images and texts created by actors in a variety of fields, from photography to books for the general public, or even for young audiences. We can see the gradual reintegration of elements of our natural environment into daily perceptions within western societies, with the stated objective of raising awareness of both the beauty and the vulnerability of our planet.

Thus, plant elements largely became visible to a receptive audience through major exhibitions and photo catalogs such as that of Yann Arthus-Bertrand (1999), based on an aesthetic approach, and designed to evoke an emotional response. However, at the same time, this aesthetic approach paints the world as a decorative element – a disembodied postcard – which is not conducive to sensory, living perception, creating the relationship with our environment. Nearly 20 years later, other approaches were developed to continue raising awareness among our peers. The forester Peter Wohlleben (2017) sought to clarify our understanding of the plants – notably trees –

Chapter written by Amélie COULBAUT-LAZZARINI.

around us, also drawing on aesthetically attractive photographs, used to support the explanatory discourse in his books.

On a scientific level, the questions of emotions (Kohlmann 2017), and perceptions, involved in considering the living world around us (Waal and Paloméra 2011; Kohn et al. 2017; Paillard and Lapierre 2022) have developed. At the same time, work in linguistics disseminated the advancing knowledge about plant communication – i.e. not only communication *about* plants, but communication *with* and *among* plants too (Gagliano et al. 2017; Kohn et al. 2017) – and helped that knowledge take root in the human and social sciences. We also witnessed the development of various practices such as forest bathing (Monce 2018; Cosquer 2021), which opened the door to plants being considered relational beings.

What is of interest to us here, however, is how these issues have been handled by public stakeholders. Indeed, when these subjects are addressed by public and regional communications (Bessières 2016), they gain validity as a matter for public interest. This creates space for thinking about how our relationship with the world has shifted. This point seems all the more important given that research is pointing to a decline in our experience with nature (Prévot 2015) and the need to "create new social models in which nature is at the center" (Prévot 2020).

This chapter highlights the way in which plants are increasingly being used in communication as a tool by a number of public-sector actors – notably, in upland territories: Regional Natural Parks, agglomerations or regional authorities.

This focus leads us to analyze how the support of territorial bodies, such as France's *Parc naturels régionaux* (Regional Natural Parks, PNR) or urban developments, for programs integrating artistic, bodily aspects in connection with the landscape, shows a shift in the trend towards plants being taken into account more fully in humans' relationship with the world. More specifically, we are interested in the way in which these programs show a relationship with the plant world from a perspective that is not purely utilitarian, involving the exploitation of an available resource, but rather one of communication – of a link to the living world, which embodies (Martin-Juchat 2008) our own belonging to the world.

4.2. Environmental communication as an anchor point in thinking about our place in the world: a shift towards relatedness

From a theoretical point of view, to consider the idea of changing from a relationship of exploitation of the world to one of reliance upon it is to accept the idea that the world does not revolve around humans. It means moving from a logic of domination, which legitimizes the exploitation of the environment as a resource at humans' disposal, to a more relational approach. While it is not new, this type of approach has not yet been widely adopted in contemporary western societies; indeed, in an attempt to delegitimize the scientifically founded idea, opponents often decry it as naivety or activism. However, numerous actors, such as those mentioned above, and many others in recent decades (Bateson 1972; Morin 1990; Latour 2010), have devoted research to it that has been widely shared, even outside of academic circles.

Grounding our observations in environmental communication (Catellani et al. 2019; Coulbaut-Lazzarini and Couston 2021), we will refer to authors from outside that field, but whose work is relevant to an enriched understanding.

Thus, it would seem helpful to point out that one of the major roles of communication is to allow us to think about the link, the relationship with the "other", in order to establish diplomatic cohabitation with the living world (Morizot 2017). Indeed, in order to achieve such cohabitation, it is essential to take account of and accept the other party's presence.

As Babou (2017) points out, in order to think about that relationship, we need to go beyond the dichotomy of nature and culture, to construct a harmonious relationship between human societies and natural environments. Elements other than humans can then exist without automatically being kept at arms' length, and without a fundamental opposition. Looking beyond that which sets us apart, we need instead to look at what binds us together.

Once we break free from this dualism, it becomes possible to move past a human-centric mindset, and think of humans as merely one living entity among so many others (Grusin 2015). This stance leaves behind the logics of anthropocentric domination, and moves towards more symbiotic relationships (in the sense in which it is meant by scientific literature, in which the concept includes only relationships involving reciprocal contributions).

From this perspective, information and communication sciences, notably in the field of environmental communication (Pascual Espuny 2008), offer a space in which we can think about connections with nonhuman entities, and in doing so, shift the paradigm in which we operate (Couston 2005, 2019). This framework opens up the question of the integration of values into a world view whereby humans are encouraged to think of themselves as being connected to "a biospheric environment recognized as a metasystem" (Couston 2019).

The overlap between these various theoretical contributions leads us to consider the relationship with the natural environment as the crucible of reliance (i.e. relatedness) (Bolle de Bal 2003; Morin and Wolton 2011; Coulbaut 2022), perceived as being a catalyst for evolution, able to construct relationships which connect humans and non-humans. In line with our earlier work, we consider "reliance" to be that triply interdependent relationship between ourselves, others and the world, in which the question of intercorporality and the dimension of imaginaries interface with each of the other components in this Web of reliance.

4.3. Methods and fields

In terms of the research methodology, our work is based on a set of interconnecting methods. Firstly, we compiled a corpus of documents produced by territorial actors in our area of study, notably including an "info-mag" published by the regional authorities of Savoie, and the introductory pages of the website of the Bauges Massif natural park. This restricted corpus illustrated the emergence of the topic among territorial actors.

We also inventoried and analyzed a number of programs supported by regional natural parks in the Alpine foothills – notably those of the Bauges Massif, Chartreuse and Vercors. Then, we conducted semi-directed interviews with stakeholders involved in these territories and/or programs. Finally, we carried out participatory observations of certain programs[1].

4.4. Plants: a crucial component in our relationship with the world

The stance of the territorial public actors and natural parks, and their effort to support projects which involve human–plant connection, help us to establish in the public space a relationship with the world which takes account of these connections. However, there are often remaining points of ambivalence, giving plants a shared role in creating more than just a shared space.

4.4.1. *Plants and territorial communication*

In Savoie, local actors such as urban communities or the regional authorities produce magazines addressing the residents of the territory, sometimes including the explicit aim of raising awareness about the local environment (see Figure 4.1).

The concept of reconnection suggested in the legend of this large illustration highlights this aspect. Proximity is also emphasized, as a factor that facilitates the creation of such a relationship, and vegetation is everywhere in the image. However, the concept of heritage, and the wooden walkway which invites readers to roam through the forest without ever actually coming into direct contact with it, demonstrate that there is still a long way to go in order to truly leave behind the dualism between nature and culture, and build a real connection. In addition, we cannot ignore the aspect of territorial marketing, showing a socially acceptable image of nature that humans can walk through, while still remaining in a carefully controlled environment (Lozato-Giotart 2008).

1 This included 10 days of observation of the eco-volunteer program run by the PNR of the Bauges Massif, as part of the project COMHUNENVI, funded by the labex ITTEM (Innovations and Territorial Transitions in the Mountains), the Zone Atelier Alpes and Fédération de recherche en environnement.

Figure 4.1. *Savoiemag, Spring 2022 edition. Caption reads: Want to (re)connect with nature? You're in luck! In Savoie, nature is everywhere; it's right on your doorstep. So, with a Spring in your step, get out there and explore our wonderful natural heritage, which is also home to a breathtaking array of biodiversity. For a color version of this figure, see www.iste.co.uk/souillard/emerging.zip*

Other elements, such as this page introducing the PNR of the Bauges Massif (see Box 4.1), reveal the desire to be taken over by the pre-Alpine territories and the vegetation to which they are home.

> "After walking through the forests of beech and fir, then spruce, the mountain pastures of tall grasses where herds of cattle graze during the summer, the steep grass where the cattle have given way to chamois goats and bighorn sheep, a few rocky outcrops sheltering lamb's-ear and edelweiss here and there, you find yourself at the summit, with a commanding view of the landscape. Every season brings new colors and a unique atmosphere."

Box 4.1. *Extract from the website of the PNR des Bauges (2022)*

It is of little consequence here (Box 4.1) that the "wild" space contains areas that have been largely shaped by pastoral activities, such as pastures. What is being promoted here is connected, more, to the relationship with the living world and the various species (plant species, in particular) that visitors to these territories can see. The image of the "tall grasses" in the mountain pastures helps us to create that feeling of a relationship, as does the picture of "steep grass" (neither phrase actually specifies which species can be found there), and the forests of different species of tree. Note, here, that the chosen plant species play on two mechanisms: familiarity with the species of tree mentioned (beech, fir and spruce), and the sense of wonder attached to rare and iconic species (edelweiss), or ones which are inextricably linked to mountain life (lamb's-ear). The coupling of these two mechanisms helps us create the idea of the pre-Alpine vegetation as elements that are both near and desirable, or capturing the imagination, with which it becomes desirable to interact and have a relationship, of sorts.

In addition, the way in which the spaces are presented, beyond the "postcard" aspect[2], attaches great importance to the world of the senses. The diversity of elements that are mentioned is used to create the image of a living world that the walker is part of, and which they are exhorted to feel and to appreciate. The introductory pages of the websites for various pre-Alpine PNRs make use of similar elements and strategies. The description and exaltation of the territory, and the invitation to come and explore it, are interwoven, together forming an initial phase, which is raising awareness. However, the role of these public stakeholders, tied up with their goals, cannot be limited to digital communication: their agendas must include real-world operations, in the field, which reflect their ambitions.

4.4.2. *Public actors creating media coverage of territories with multiple issues at stake*

Therefore, we need to look at how these actors use the idea of a relationship with the plant world, in communication for the purpose of creating an effective link, sensory "diplomatic cohabitation" (Morizot 2017), with the living world.

That same sensory universe is put to use, beyond the digital communications of regional public actors, by the implementation of various types of actions. A possible variant can be observed in the rollout of field activities integrated with the missions and skills of the PNRs. The mediation implemented by the ecovolunteers of the park at the Creusates peat bog is a

2 While we cannot ignore the illustration using a "beauty shot", and the techniques of territorial marketing (Chamard and Schlenker 2017) to focus readers' attention on the idyllic, archetypal aspects, and concealing the more controversial aspects (Latour and Hermant 1998), it is the presence of the sensory dimension which is of interest to us in this article.

prime example. It is human communication, with support in the form of "pedagogical kits" containing books, handling objects and other didactic materials. The mediators introduce, explain and demonstrate the evolution of the vegetation growing in the bog. By allowing the audience to observe an explanatory timeline on the ground, along with publications and models, and then to touch, see and smell the environment in which they are walking (Chambru 2021), the media creators enable the audience to form a direct, concrete link with the plant world.

In addition, there are projects, residencies for research and creation (Blanc and Legrand 2019), funded as part of programs supported regionally or nationally. Thus, in the content of the agreement (2018–2020) governing relations between the DRAC Auvergne-Rhône-Alpes and the Parcs naturels régionaux, hybrid residencies[3] bring researchers in social sciences and artists together. In 2021, an anthropologist and a choreographer were selected, offering audiences the chance, for example, to rediscover a section of the trail known as "Au fil de la Bourne" (running along the Bourne Gorge), by experiencing dance gestures[4], inviting them to take better note of the richness of the environment (including the sounds and atmosphere of the prairies and forests). The choices made by public bodies in terms of calls for projects and the selection of such projects attest to the integration, to some extent, of this idea of rapprochement between the human and plant worlds. While experiments uniting arts and sciences are certainly not a new phenomenon, what is interesting here is that they are supported by public authorities through calls for projects at the regional level – no longer merely at the microlocal level. Beyond practices which are well established but few and far between, the recent renewal of a multi-year regional agreement is indicative of the strengthening of public attention.

Through these various devices, we see the foundations of acceptance of the sensory framework of our environment, and a step towards a relationship with the world that, little by little, is including the construction of a relationship between humans and the plants surrounding us.

However, we cannot take account of only those aspects which support this interpretation, and it must be recognized that there are numerous

3 See the call for projects: https://www.culture.gouv.fr/Media/Regions/Drac-Auvergne-Rhone-Alpes/Files/appel-a-projets/Appel-a-candidature-residence-chercheurs-et-artistes-itinerance-culturelle-dans-les-PNR.

4 See: https://www.lolink.net/ciel-ouvert.

programs based on conventional modern thinking, which are supported by the same public actors. An example is the *Journées forêts ouvertes* (Open Forest Days), to "show you in the flesh the numerous services that the forest provides [...] Producing wood to build, furnish and heat your homes [...] Protecting your water resources"[5]. While these days are organized by a private forestry group, it is supported by the Conseil Savoie–Mont-Blanc (formed by the merging of the regional councils for Savoie and Haute-Savoie). It seems that far from being a realized utopia, the transition towards a paradigm shift is still at an embryonic stage.

4.5. Connections with the plant world: towards reliance?

The way in which these programs allow connections to be forged with the plant world suggests a shift in the understanding of the living world, and the gradual acceptance of possibilities for communication. The proposed mediation systems represent an attempt, by connecting to ourselves, to others and to the world around us, to move away from the rhetoric of a world which belongs to humans, towards a perception of a world to which humans belong – of which they are intrinsically part. However, we can see a number of instances of ambivalence and incompleteness which plague both the communication by public actors and the choice of events and projects that they support. While changes are coming about, in order to have these systems cooperating, it is necessary to have a reliance-based approach, which has yet to be constructed.

This construction should go beyond the simple analysis of the systems which do or do not contribute to reliance, making it effective. This could involve greater closeness with the media creators involved and capable of feeding into the reliance, as well as by working with their audiences, without forgetting the element of time which interferes with these different actors.

4.6. References

Arthus-Bertrand, Y. (1999). *La terre vue du ciel. Un portrait de la planète en l'an 2000*. La Martinière, Paris.

Babou, I. (2017). L'atelier politique de la nature. *Questions de communication*, 32(2), 7–28.

5 See: https://www.ufp74.fr/event/forets-ouvertes-a-saint-andre-de-boege-2/.

Bateson, G. (1972). *Vers une écologie de l'esprit, tome 2*. Points, Paris.

Bessières, D. (2016). La communication publique environnementale : un enjeu de développement durable pluriel et contingent. In *La communication environnementale*, Libaert, T. (ed.). CNRS Éditions, Paris, 219–232. doi: 10.4000/books.editionscnrs.20943.

Blanc, N. and Legrand, M. (2019). Vers une recherche-création. Textes, corps, environnements : explorer la portée transformatrice des récits dans les relations au milieu de vie. *ACME: An International Journal for Critical Geographies*, 18(1), 1.

Bolle de Bal, M. (2003). Reliance, déliance, liance : émergence de trois notions sociologiques. *Sociétés*, 80(2), 99–131.

Catellani, A., Pascual Espuny, C., Malibabo Lavu, P., Jalenques Vigouroux, B. (2019). Les recherches en communication environnementale. État des lieux. *Communication. Information médias théories pratiques*, 36(2). doi: 10/ggq49b.

Chamard, C. and Schlenker, L. (2017). La place du marketing territorial dans le processus de transformation territoriale. *Gestion et management public*, 6/1(3), 41–57. doi: 10.3917/gmp.061.0041.

Chambru, M. (2021). L'alpage du Champet : arpenter la montagne pour repenser la médiation scientifique. *Journal of Alpine Research, Revue de géographie alpine*. doi: 10.4000/rga.8708.

Cosquer, A. (2021). La sylvothérapie [Online]. Available at: https://www.librairie-dedicaces.fr/livre/9782715403963-la-sylvotherapie-alix-cosquer/.

Coulbaut, A. (2022). Coopération et reliance : design et communication environnementale comme logiques de pensée et d'action au service d'enjeux complexes. *Approches Théoriques en Information-Communication (ATIC)*, 4(1), 127–137. doi: 10.3917/atic.004.0127.

Coulbaut-Lazzarini, A. and Couston, F. (2021). Manifeste pour penser la communication environnementale à la lumière des liens humains/non-humains. *Revue française des sciences de l'information et de la communication*, 21. doi: 10.4000/rfsic.10021.

Couston, F. (2005). *L'écologisme est-il un humanisme ?* L'Harmattan, Paris.

Couston, F. (2019). Les SIC : une discipline ouverte sur le nouveau paradigme écologique-complexe ? *Sciences de la société*, 104. doi: 10.4000/sds.11208.

Gagliano, M., Ryan, J.C., Vieira, P. (2017). *The Language of Plants: Science, Philosophy, Literature*. University of Minnesota Press, Minneapolis [Online]. Available at: https://www.jstor.org/stable/10.5749/j.ctt1nxqpqk.

Grusin, R. (ed.) (2015). *The Nonhuman Turn*. University of Minnesota Press, Minneapolis.

Kohlmann, É. (2017). Gérer/aimer. Communication et nature dans le parc naturel régional du Pilat. *Questions de communication*, 32, 51–74. doi: 10.4000/questionsdecommunication.11432.

Kohn, E., Descola, P., Delaplace, G. (2017). *Comment pensent les forêts. Vers une anthropologie au-delà de l'humain*. Zones Sensibles, Brussels.

Latour, B. (2010). *Nous n'avons jamais été modernes. Essai d'anthropologie symétrique*. La Découverte, Paris.

Latour, B. and Hermant, E. (1998). *Paris ville invisible*. Empêcheurs de penser rond, Paris.

Lozato-Giotart, J.-P. (2008). *Géographie du tourisme. De l'espace consommé à l'espace maîtrisé*, 2nd edition. Pearson Education France, Paris.

Martin-Juchat, F. (2008). Penser le corps affectif comme un média. *Corps : Revue Interdisciplinaire*, 4, 85. doi: 10.3917/corp.004.0085.

Monce, L. (2018). *Ces arbres qui nous veulent du bien – À la découverte des bienfaits de la sylvothérapie et du bain de forêt*. Dunod, Paris.

Morin, E. (1990). *Introduction à la pensée complexe*. ESF Éditeur, Montrouge.

Morin, E. and Wolton, D. (2011). Grand entretien. *Hermès, La Revue*, 60, 241–254.

Morizot, B. (2017). Nouvelles alliances avec la terre. Une cohabitation diplomatique avec le vivant. *Tracés. Revue de Sciences humaines*, 33. doi: 10.4000/traces.7001.

Paillard, B. and Lapierre, N. (2022). Les vivants dans tous leurs débats. *Communications*, 110(1), 5–27. doi: 10.3917/commu.110.0005.

Pascual Espuny, C. (2008). Développement durable : les SIC, pertinentes et sous-estimée. In *Congrès SFSIC Compiègne 2008 – Les sciences de l'information et communication, émancipation et pluralité*, Compiègne [Online]. Available at: https://www.sfsic.org/evenements-sfsic/congres-sfsic/congres-sfsic-compiegne-2008-les-sciences-de-linformation-et-communication-emancipation-et-pluralite/congres-sfsic-compiegne-2008-axe-les-origines-des-sic/.

Prévot, A.-C. (2015). Se mobiliser contre l'extinction d'expérience de nature. *Espaces naturels*, 51 [Online]. Available at: http://www.espaces-naturels.info/se-mobiliser-contre-extinction-experience-nature.

Prévot, A.-C. (2020). De nouvelles relations à la nature pour des changements transformatifs de nos modèles de société ? *Annales des Mines – Responsabilité et environnement*, 100(4), 23–27. doi: 10.3917/re1.100.0023.

Waal, F.D. and de Paloméra, M.-F. (2011). *L'âge de l'empathie. Leçons de la nature pour une société solidaire*. Actes Sud, Arles.

Wohlleben, P. (2017). *La vie secrète des arbres*. Les Arènes, Paris.

5

Nature in Abandoned Ruins: A Resource for Environmental Education

5.1. Introduction: opening our eyes to abandoned ruins

In recent years, there has been recovering interest in abandoned modern ruins, where uncontrolled vegetation emerges from ground previously covered by a now-abandoned artificial construction. For example, the ruins of the Palais d'Orsay, a late-19th-century wasteland at the heart of Paris, have been described as "an extraordinary garden in Paris, where absolutely fabulous, even exotic, vegetation grew: all the seeds from the greenhouses of the neighboring private hotels had blown into the Courtyard, giving birth to the most fantastical plants" (Crosnier-Leconte 1986). Between the realms of cultural heritage (in which ruins are viewed as being precious traces of the past) and natural heritage (in which they are viewed as meaningless), even today urban development leaves spaces similar to the Palais d'Orsay (Wadbled 2022): wastelands, in which vegetation is free to develop. Alongside developed spaces that are open for various uses, these spaces in which vegetation freely proliferates are often hidden from view, or forbidden.

Thus, when we visit these places, we are inevitably struck by the recovering wildness (DeSilvey 2017), and we find vegetation that is neither wild nor entirely domestic. How does wildness become a unique type of existence, shared between the ruined buildings and the recovering vegetation, which come together to constitute an abandoned ruin? How do we build a relationship with abandoned ruins, and in this context, how can we perceive a unique landscape? What educational value could such places have

---

Chapter written by Nathanaël WADBLED.

in the context of building our relationship with the environment? These questions will be examined with an epistemological approach, the aim being to explore how a set of forms and materials become a recognizable landscape, and constitute what visitors perceive as the reality of their surroundings. Thus, this is a preliminary work, presenting frameworks we can use to examine ruins where wildness is recovering – as opposed to a more detailed field study, which would be based on a corpus of sites, interviews with visitors and an exhaustive bibliography of representations of such ruins.

5.2. Communication of wild ruins

5.2.1. *The environmental form of heritage*

In order to formally describe ruins in the process of returning to the wild state, we need to recognize an environment where nature freely develops in the remains of human constructions. At first glance, a site contains an assortment of materials and vegetation. When we pay close attention, within this unique environment, we can recognize two facets: wilded artificiality and feral nature. While the concept of something "feral", strictly speaking, refers to a domesticated species that has returned to the wild state, it more generally refers to a situation where plants or animals develop without human control, though originally, their very existence may be due to human activities. In this particular instance, the plants and animals which make their home in abandoned ruins would not exist, as such, had humans not artificially built on the site in the first place (Maris 2018; Schnitzler and Genot 2012). The two notions of wilded artificiality and feral nature come together to form the notion of Anthropocene heritage (Wadbled 2021), which includes humans' relationship with the environment to which they belong.

This form of heritage carries a certain degree of dissonance, because it is composed of traces left by a multitude of agents, all of which could rightfully claim that the heritage is theirs – humans among them. Indeed, the material forms of ruins, in the process of returning to the wild state, are the remains of human buildings and developments, which were built for the purpose of some cultural activity. While it is possible to identify what those intended uses were, based on the remains, through knowledge of a site's history or the history of technology, the heritage which it has become is plainly not material culture designed to preserve the memory of those human activities. In being abandoned, the forms have changed over time and fallen into a state of ruin, which itself carries the heritage of multiple acts of weather, geology, animals and vegetation. Indeed, the fact of being abandoned means that the constructions are no longer protected from these elements, as are buildings which are maintained. From the standpoint of the human group which once occupied the site, or of a historian studying social habits of the past, that transformation is a decay. From the perspective of the life forms which have made a home there, it is the recovery of a living environment. Ruins returning to the wild state, then, are a space similar to those in the Amazon rainforest where anthropologist Eduardo Kohn recognizes that "history, which remains embedded in the forest, is mediated by a form which alters it, and which cannot simply be boiled down to events or human environments" (Kohn 2017, p. 317).

In today's cultural context, in order to engage with the remains of the past, we must recognize their value as a historical artifact; this is impossible to do without historical awareness, which must be inculcated (Poulot 1997). On the other hand, there is a moral and artistic dimension to the sense of nature, which is viewed as arising and growing only in protected, unspoiled spaces (Chalvet 2021; Luglia 2021). This being the case, the environments of interest to us here, where the artificial and the natural collide, do not really hold a social value. Firstly, they do not carry enough historical significance to be of interest, and secondly, the nature that is found there is too polluted to be valued. These spaces are not pieces of historical heritage, protected from nature's onslaught in order to preserve a memory worth passing down to future generations (Davallon 2006). On the other hand, neither are abandoned ruins "natural" spaces per se; "natural" spaces are viewed as primitive, and thought of as representing an environment that has been sheltered from significant alteration by humans (Callicott 2008). This division of heritage into cultural and natural is part of the general framework of what natural anthropology calls naturalism, which "is our own mode of identification and permeates our shared meaning as well as our scientific practice" (Descola 1996, p. 66). Under this system, the world is composed of a) an artificial domain which is constructed and organized by humans on the basis of cultural preferences and b) a natural kingdom in which beings are subject to the laws of nature. If these domains overlap in a historical heritage site, then the cultural heritage can be seen as being degraded, and the natural heritage as being polluted.

As this dichotomy between the two categories of ruins makes no allowance for the Anthropocene, no value is attached to abandoned ruins. Wild ruins come into being when a space is no longer viewed as being worthy of interest: worth maintaining and protecting from nature. It then quickly becomes a forbidden space, due to the dangers posed by the buildings' decaying structures. Hence, wild ruins are not a part of most people's day-to-day experience of life. In fact, often, it is not specifically stated that they are uninteresting, because the signage gives no reason why access is forbidden. To an outside observer, these sites often seem to be private properties like any other.

In order to recognize Anthropocene heritage, we need to attach it to a category other than those defined by a naturalist ontology. Hence, the definition of what is natural and what is artificial must change. Artificial constructions where nature has been tamed, and the wild world where the artificial has no impact at all, are no longer viewed as two discrete and mutually opposing states: they are defined as the theoretical extremes of a hybrid reality in which the various conditions constitute a continuum (Lesslie and Taylor 1985). Wildness is not a threshold level of environmental impact below which nature remains wild. It is defined from the standpoint of the beings which inhabit the environment, rather than solely from the standpoint of the organization of human society. These inhabitants include any visitors to abandoned ruins, who need to find their place within an environment that they share with others. Consequently, the epic significance of the effort invested in constructing what used to be a functional building, creating an environment liberated from the constraints of nature, is lost. Instead, the labor is perceived only as a temporary effort, to keep wildness at bay, prior to the site's abandonment. In parallel, fencing off abandoned ruins is not a way of preserving a natural space, but rather of preventing the spread of wildness (which is the normal condition in which an artificial environment should be).

The anthropology of nature (Kohn 2017) and critical heritage studies (Harrison and Sterling 2020) identify situations where we see such an ontological shift in the practices of societies whose relationship with nature is not based on a dichotomy between the natural and social worlds. Indeed, such societies have social practices which encourage people to form relationships with nature – in the same way as a naturalist society has institutions such as heritage conservators to keep nature and society strictly separate. This is an epistemological stance which rejects the exoticism implied by the opposition between ourselves, removed from nature and others who remain within it (Bensa 2006). In line with such a stance, paying attention to abandoned ruins at the heart of the territory of contemporary societies offers the opportunity to break the habit of creating a dichotomy between cultural and natural heritage, without having to use – and thus instrumentalize – practices imported from elsewhere. Paying attention to ruins returning to a wild state is part of an approach which calls both the effectiveness and the hegemony of the naturalist paradigm into question. Though naturalism is recognized as being characteristic of modern society, the many practices in which it is tested represent situations in which hybrid relationships are constructed – for example, in the context of hunting (Stépanoff 2021).

5.2.2. *Landscape perception*

As abandoned ruins are often closed to the public, or difficult to access, photographs taken by visitors to document their experience in ruins where nature is making a comeback constitute a prime point of observation for how visitors respond to what they are seeing (Veillon 2021; Jonk 2022). Such photos are taken in the practice of urban exploration (urbex), which involves visiting forbidden spaces that we do not see as part of our day-to-day social lives (Offenstadt 2022). While this practice feeds a taste for rule-breaking (Garret 2012) and interest in the history of that which official heritage has forgotten (Offenstadt 2018), photographs of abandoned ruins show a specific situation in which abandonment leads to the return of wildness (Wadbled 2020). These images are explicitly presented as recording the photographers' experiences when visiting these places, and are recognized by urbex lovers as being reminiscent of their own practices. By showing abandoned ruins, they suggest that the visitors have been touched by the aesthetic beauty of spaces in which wild nature overlaps with human structure.

These images are neither the simple record of what visitors have perceived, nor the manifestation of artistic choices stemming from the photographer's tastes or sensibilities. The choice of framing, field depth and color palette expresses the photographer's response to what is being shown. It frames their experiences of the landscape, which emerge through a coming together when the visitor connects with the site, on a sensory and emotional level (Droz and Miéville-Ott 2005; Bigando 2012). In pragmatic terms, the perception of a landscape on a site can be compared to the aesthetic visual effect produced when we read a book: identifying the elements which are present, and recognizing their arrangement in a coherent whole (Iser 1976). Thus, urbex photographs can be approached in the way in which anthropologist Philippe Descola suggests that artistic images in general should be considered. He is interested in how they form an "active trace not only of the objects they show, and with which they share recognizable properties, but also of the essence of those objects", i.e. the "ontological framework of reality" (Descola 2021, p. 31). It takes form in geographic imaginaries (Debarbieux 1995), i.e. ways of organizing the landscape in which objects and beings appear so that they are recognizable.

The landscapes captured in this way are often naturalistic. On the continuum of possible interactions between what is wild and what is artificial, a particular value is attached to the opposing poles, at which we dominate the other: wild nature returning to win through despite human impact; or nature destroyed by industrial modernity, where not even humans are able to live any more. Either human constructions have no significant impact on wild nature, which returns as though nothing had happened; or pollution makes it impossible for life to flourish, and it is reduced to a poor and poorly form (Wadbled 2021, 2022). This duality is reminiscent of the distinction between cultural and natural heritage: the artificial and the wild reign in two different realms. A space's habitability does not depend on the degree to which they coexist. There are only two possible forms, depending on whether human pollution or wild nature wins out to dominate over the other. The Anthropocene that is depicted is not an environment which is at risk of total destruction, or which can be given back to nature. It is the alarming shift in ecological dynamics in our modern world which is driving us headlong into disaster (Bonneuil and Fressoz 2013), unless we can still find some way to arrest the process. The only alternatives in the face of the apparently inexorable decay of human constructions seem to be a primitivist acceptance of nature's dominion (Denogent 2022); or a heuristics of fear of a world in which life is impossible – or at the very least, intolerable (Jonas 1991).

5.2.3. *The relevance of the available geographic imaginaries*

To determine the relative pertinence of viewing abandoned ruins as being feral or naturalist, the usual way of critiquing an idea, by pointing out a contradiction or overinterpretation, is not relevant. Since by definition, abandoned ruins were not intended to be so by their creator, the most relevant imaginary does not stem from the meaning that they held previously. Unlike parks or heritage sites which are organized with a particular goal in mind, ruins do not have an enunciative aspect. To choose one interpretation over another to represent what wild ruins truly are, we must adopt one or other of the two approaches: either we consider that an object's form is defined and utterly immutable, or we accept that it is defined by people's perception of it.

To give both approaches an equal opportunity, we can qualify them as specific types of what sociologist Alfred Schütz (1998) called "relevance systems" – i.e. they are the relevance systems of the semiologist and of the photographer who engage with the ruins. In a phenomenological approach, Schütz argues that the meaning we attach to experiences does not depend on the nature of the objects, but on the way in which they are interpreted in the observer's a priori schema. The semiologist and photographer are part of two different epistemological approaches, both of which could arguably be considered relevant: inductive and deductive. The former describes what is perceived, and from this observation, pragmatically deduces the most relevant imaginary. The latter, meanwhile, produces an a priori imaginary, associated with a desirable experience, and critiques what the observer is seeing to assess whether the site allows for that experience. The difference between the two is not their epistemological value. They are two dynamics of knowledge which give primacy, respectively, to the search for what is most homogeneous about the site as it is in its present state, and to the tradition which assesses experience from the perspective of shared meaning.

Firstly, with semiotic analysis, a landscape is recognized on the basis of the most complete and rigorous possible examination of the site. It consists of a description which marks how meaning can emerge from the arrangement of the various visible elements. If meaning is deduced by the searcher, it constitutes an interpretation of the site, rather than the projection of the searcher's own viewpoint. Each singular element makes sense as a figurative element because of its particular position within the whole arrangement which makes up an immanent structure. If the imaginary deemed relevant is part of the stock of those which make sense to the searcher, its relevance is established on the basis of the principles of completeness and parsimony that are characteristic of the scientific method: the interpretation produces a landscape which integrates all the elements present and arranges them without imposing any external principle of order upon them. From this standpoint, recognizing a wild ruin as being feral is a pertinent interpretation, because identifying it as a decayed or polluted ruin fails to do justice to its specificity in relation to historical heritage and wild nature.

Secondly, the landscapes shown in the photographs taken on urbex expeditions are not the result of a rigorous interpretation of the available content, but rather, its integration into the photographer's own field of experience. Thus, the geographic imaginaries of wildness and pollution reinvest the topoi of the history of art, notably ingrained in romantic and post-apocalyptic art (Wadbled 2020). If urbex photographs are indeed a portrayal, or figuration, of the experience of visiting the site, rather than purely an artistic choice, then the ways in which these ruins are portrayed constitute the frameworks which the photographers use to derive meaning from what they see (Roger 1997; Staszak 2006). Figuration is a projection of the photographer's culture. Provided what is perceived can fit into this framework, its heuristic relevance is confirmed. Provided others recognize that landscape and are therefore working with a similar framework, its social relevance is strengthened. Thus, it is constructed in relation to a shared sense of meaning, rather than on the basis of the individual's interpretation of the site.

Rather than laughably rehashing the epistemological division between realism and constructivism, we can recognize, in that division, the tension between two individual visitors' differing interpretations of the landscape. The visitors' relationship with the site can, indeed, be broken down into a) how they are touched by the surroundings and b) how they respond to that stimulus, seeking to interpret it. The most relevant landscape to describe what is being perceived, therefore, may not be consistent with the available categories of landscape. This causes a break in the experiential dynamic recognized by phenomenological sociology as "rendering familiar that which is topically relevant, by searching in the visitor's stock of experiences for elements which are 'reminiscent of', or at least which are similar to, the subject at hand" (Laoureux 2008). The visitor's response comes pragmatically in the form of a translation: when viewed through the lens of the available frameworks, the landscape is transformed. When we set out to understand the experience of visiting these sites, we must recognize that it cannot be reduced to the visitors' awareness of their own point of view, and we must also recognize that this perspective depends on the interplay between the site, as it appears in reality, and the visitors' stock of available imaginaries with which to interpret it. Thus, a description of the experience of visiting the site would be at once a pragmatic anthology of the ritual of attention to the effect of an activity, independently of the frameworks within which the visitors interpret it (Houseman and Severi 2009), and a sociological examination of memory, tracking the process by which the representation of the experience takes shape (Halbwachs 1994, 1997).

5.3. Capitalizing on a communicational potential

5.3.1. *Interpreting indeterminate spaces*

If, when visiting a place, we feel a tension between feral nature and its naturalist representation, the visit does not prompt the visitor to enter into a relationship with the feral landscape. For such a site to become a resource for environmental communication, some guidance is needed. This involves the work of a mediator – a guide – whose job is to forge a relationship between the visitors and the site. Thus, their role is not to explain what is perceived to be most relevant about the site, but instead to shed light on the process whereby each individual perceives a landscape, and to challenge what can be spontaneously considered a natural attitude. Thus, the approach to the site takes place within an epistemological framework which historian Carlo Ginzburg (1980) describes as an evidential paradigm: recognizing a material object as being evidence of nature's return or destruction is believed to attest to the observer's own reality. This meaning is not recognized as being projected on the basis of the observer's own stock of a priori representations. Thus, participants are led to a pragmatic attitude, attentive to the dynamic of relevance structures. The site appears as what the literary theorist Wolfgang Iser calls a "place of textual indeterminacy" (Iser 1976, p. 52), which forces visitors to question the relevance of their existing imaginaries: the experience of the landscape is then akin to a search for meaning, constructed through a series of complex interactions between different imaginaries and the lived experience. Philosopher William James describes the effect produced by this epistemic clash as "the effort necessary to accept novelty while presenting it in as familiar a way as possible" (James 2007, p. 124).

This attitude echoes that of critical pedagogy based on an "action-reflection approach" (Pereira 2017, p. 110), rather than that of edification, in which an external truth is imposed. While it retains the desire to demystify visitors' natural relationships with the landscape, by asking why things are the way they are, it sits apart from criticism, social sciences and pedagogy because there is no normative basis (De Munck 2023). If, as a starting point for our thinking, we consider the complexity of a relationship with the site, it problematizes the visitors' sensory response to the site rather than the intrinsic quality of the geographic imaginaries. Critique, then, reflects on the limited relevance of using the available categories to interpret the site. It targets imaginaries which define an image of a ruin which is coherent, but does not correspond to the site as it stands. Critical thinking would assess the different imaginaries on the basis of extrinsic criteria – notably political and moral – of which visitors would be unaware, due to insufficient education. For example, to people who are fully aware of the ecological meltdown currently in progress, a landscape which does not reflect the catastrophic damage wrought upon the natural environment by modern civilization would appear to be false knowledge, or untruthful. This is the type of guidance which is traditionally in place on historic or natural heritage sites, the aim being to inculcate a representation of nature or of history that adheres to certain principles of edification, rather than to highlight historiographic controversies (Löwenthal 1998). In order to construct this new and true framework, the false ones which uninformed visitors might have held previously have to be broken down; the two processes are inextricably linked. The purpose of bringing visitors to the site is to replace their false narratives with true ones. This is done by creating awareness, through the guide's explanations, raising visitors' awareness and guiding them towards the appropriate attitude.

5.3.2. *Mediation in which the "visitors play a starring role"*

This approach takes place in parallel to lectures and performances which trigger reflection about shifting our representations of nature (Aït-Touati and Penot-Lacassagne 2022). What these approaches have in common is the desire to open the door to a new experience of humans' place in the world. In both cases, they are performances, in that their value depends not on the quality of the discourse, but on the effect produced when the receivers invest in the communication process with their own experiences. The unique aspect of visiting wild ruins relates to the mediation system, which involves both conscious and corporal experiences, but in different ways. With spectators in a theater, their attention is commanded by and their reflection is shaped by what is happening on stage. Visitors being led by a guide to a real-world site, on the other hand, are standing, mobile, walking through an immersive environment which stimulates their senses. The visitors' very bodies are engaged in a material space, with which they have a bodily interaction. The emotions and reflections triggered by the performance are processed as part of a relationship which is both sensory and reflective.

In this situation, the guide may follow the principles of interpretation and guidance that apply in the context of science museums (Schiele 2001): a guide directs the viewers' attention, and explains the dynamics involved in experiencing the expographic systems. These systems create an atmosphere which is "expressed in terms of sensations, impressions, feelings, for which there are no words" (Balaën 2002, p. 88). The mediator does not produce their own discourse, but guides visitors on how to have an aesthetic experience which involves relating and responding to the site, rather than merely viewing it as an object. The practice of a guided visit, then, is radically different from that which we generally see at heritage sites (guided tours). In such conventional circumstances, the space of "interpretative cooperation which involves both the content and relationships" (Gellereau 2005, p. 191) is often reduced to a discussion about the knowledge the guide has dispensed, to which the participants are invited to respond. On the other hand, accompanying visitors on a visit in order to enrich their experience involves a different communicational situation: each visitor creates their own interpretation of the site and shares it with the others, rather than absorbing the guide's discourse handed down from a position of omniscience. The participants ask questions of the guide, interact with them and challenge them. The guide's role is not to be an all-knowing conveyor, delivering a carefully curated visit targeting a given audience, but rather a translator, facilitating mutual understanding between visitors who are engaged in the experience.

In such a system, participants are first asked to describe the landscape that they perceive, and to explain the arrangement of the various elements making up that landscape. They construct a narrative surrounding their own presence in the world, in an environment where they are free to look wherever they wish, and focus their attention on whatever they choose. In this way, the site comes to life as intangible cultural heritage, through the visit and through the way in which each individual experiences their own, deeply personal relationship with the remnants from the past, and with the natural life which is once more encroaching upon the site. Each individual becomes the mediator of their own experience, interpreting the site on the basis of their own relationship with it, and sharing that interpretation. The dynamic of the visit, then, is similar to that of a choose-your-own-adventure book: it tells the story of the participants (Davin 2020), and the adventure unfolds as they come into contact with different elements of the site.

The guide's main role in this situation is not to convey information nor even to talk, but to encourage the visitors to speak, and listen to them. In visiting a site, each individual is encouraged to state their point of view, which may be challenged and possibly changed – both through critical feedback and through recognizing others' points of view. In terms of performative effect, the narrative presented by the participants sparks a discussion, retroactively stimulating a response from the guide. Using a method that was formalized in philosophical thinking workshops run with children and adolescents (Hawken 2016), the guide turns the communicational situation into a Maieutic dynamic: responding to the participants and problematizing what they say. First of all, they try to track the various imaginaries which the visitors describe. Each person takes part in an empathy exercise, stepping into the shoes of the others to appreciate the site from their point of view. By moving from one viewpoint to another in this way, it is possible to assess how fully the elicited imaginaries are able to recognize the peculiarity of the site being visited, and the different ways in which it is presented in these various imaginaries. Secondly, the guide explains the mechanisms at play in each landscape. They put these imaginaries into perspective, and into their cultural context. In this case, the purpose of conveying knowledge about cultural issues linked to the relationship with nature is not to construct intellectual capital, but to make practical use of it. The aim, at the time, is to allow everyone to perceive the uniqueness of each landscape, and appreciate its diversity. At the same time as responding to the participants, the guide is also engaged with the visit, and recounts their own experiences. They explain the way in which the feral landscape came into being, and present their own narratives alongside those put forward by the participants. They do not claim that their ideas ought to be accepted over others, by a rational actor fully aware of what is most relevant. Rather, their view is presented as being the result of the guide's own introspection about their relationship with the world. Thus, it is not a vision handed down from on high by the guide, to which the participants subscribe and, in so doing, construct a shared world view; rather, it is a suggestion. Rather than content, what the guide passes on to the participants is a method by which they can examine what they see in the landscape: a skill rather than knowledge.

All group members implement this method jointly, working together in a shared exploration whose dynamic is led by the guide. The community which is formed in such a group is knit together by the conflicts and controversies which are sustained in a lively discussion. This dynamic can only exist if the group abides by an ethics of discussion, wherein each participant holds equal status as a legitimate speaker. The implicit normative base of the visit, therefore, is not the content that is considered to be most relevant, but rather a model of mediation which sustains a dialog around shared uncertainty. Thus, the means of participation is different from that which is to be expected in visits where a specific set of content is conveyed – as is the case in exhibitions or educational guided tours. In the case of the site visit to wild ruins, everybody is actively engaged in proactively discovering a determinate meaning: they help us to construct it, rather than discovering a pre-existing meaning. Thus, together, participants go along with the guide, deliberating as to how best to interpret what they are seeing. They are the partners of a guide who brings a motivating project capable of capturing their interest, and are, so to speak, in communion with one another, with everyone ultimately leaning in the same direction. In this case, adaptation to the audience means offering each individual a way of doing this which suits them, on the basis of their dispositions, interests and cultural capital. It is a matter of optimizing the visit system so as to convey content in a way which communicates with audiences who are looking to have an experience rather than to simply learn something (Davallon et al. 2006, p. 162).

5.3.3. *Instituting a visit*

In addition to defining the principles on which it must be based, we must also be able to find a framework in which such a visit is possible. Indeed, typically, it is not legally possible to officially organize such a visit. It would involve taking responsibility for visitors in a space which they are forbidden from entering – along with the attending consequences in terms of civil and criminal liability. While they are generally not in use, abandoned public or private spaces can be rather unsafe, and if visitors were allowed to enter, in the event of an accident, the site owner would be held liable. Therefore, they can only be explored by people engaging in urban exploration, illegally.

In addition, if the visit is intended to offer a form of environmental education, then in order for their words to be heeded, the guides need to be recognized as legitimate speakers. In recognized heritage sites, the guides'/lecturers' discourse is protected by a certain status, conferred upon them by the institutions they represent and by their formal qualifications. However, urbex visits and tours of heritage sites both focus on the transmission of history; a view on relating to the landscape may be more easily recognized as being legitimate if held by people who are passionate about the subject, or local residents, sharing their experiences. Like witnesses, they describe how they experience the spaces, and their words are received as a life story. However, leading a visit requires both relational and scientific skills. Thus, were Anthropocene heritage, identified as such, to be institutionalized, then that institution could lend legitimacy to guides who are in a position to communicate something within the domain of anthropology of nature. This institutionalization could be modeled on that of urban exploration, in the form of ruins tourism, or the heritagization of rural landscapes by the *Parcs naturels régionaux*.

The development of visits to wild ruins is dependent on the ability to host such visits and institutionalize guidance, while providing an enriching experience in which visitors discover the abandoned spaces. Indeed, establishing tourism or, on a smaller scale, opening sites up to visitors (with guides, for reasons of safety), and heritagization, may appear to contradict the very principle of visits to abandoned ruins. Sociologist Hartmut Rosa (Rosa 2018) recognizes this tension, between making experiences available (and thus curating them) and retaining the possibility for an authentic relationship with the world (free from outside influence or interference). However, rather than a direct opposition, Rosa instead describes the situation as dialectic. It is not a question of institutionalizing spaces that have hitherto been informal or indeterminate, but of institutions' ability to offer an experience that opens the door to a relationship with the world that has an emotional effect on visitors, as opposed to the consumption of an activity for its social value.

Ruins tourism is the most explicit form of institutionalization of visiting abandoned ruins. It spotlights the abandoned character of wild ruins, while providing safety guarantees. Paradoxically, this means the sites cannot be left in their abandoned state (Le Gallou 2018). Abandoned heritage, then, becomes the heritagization of abandonment, through infrastructure development. While institutions highlight the value of ruins as a forgotten cultural heritage, they inevitably make the wildness of the sites into a feature of that heritage. However, rather than actually experiencing the reality of a site for themselves, what visitors experience is more a material culture of urban exploration (Binnie et al. 2007; Wadbled 2020). When seen from a signposted walkway, these ruins are themselves a signpost towards what the exploration of such spaces should be, and the relationship with nature that it should bring. Such organization means that the visitors must be positioned as spectators, viewing the nature but not actually involved with it or participating. Thus, the heritagization of abandoned ruins is practically identical to that of natural spaces: visitors are invited to experience the places in a type of reserve, where the environment appears to be the result of a deliberate conservation effort, organized in a dedicated space (Chalvet 2021; Luglia 2021). Like wild spaces, feral spaces are enclaves where the protected environment is fenced off.

In addition to these forbidden spaces, the countryside has other sites which are accessible – particularly in the *Parcs naturels régionaux* (Baron and Lajarge 2015). Indeed, people may visit wild ruins in a natural rural environment where anthropization occurs in a natural milieu, with the two together forming a dynamic environment (Locquet and Héritier 2021). While in this situation, the preservation of the landscape is a form of recognition of "natural monuments", associated with a regional or national cultural identity (Chalvet 2021), nevertheless it presents a hybrid space – one which is both cultural and natural. The development of the territory takes the form of a relationship with nature, rather than the domestication imposed upon it. Thus, it is an inhabited space where cultural practices are inseparable from a relationship with the environment. In particular, the apparently natural parts are the result of this dynamic. Indeed, anthropization has left its mark on landscapes: while many landscapes may appear natural, they are, in fact, usually rewilded, and could be considered abandoned ruins. These sites make up the majority of the landscape: species of trees which are there because of the changing ways in which humans have made use of the forest; stone walls along paths or in the undergrowth, leveled pieces of land in the middle of a forest, traces of enclosures, etc. These sites are less spectacular and more easily accessible than those associated with urban exploration. They are of little interest to urban explorers, because they do not represent defiance of a disciplinary organization of the space which prohibits access to certain areas, or the adventure of discovering exceptional sites that are difficult to access. This shift makes contact with the environment, and with the Anthropocene in the environmental sense of the word, more of a day-to-day experience. A visit to these places is part of ecotourism which develops sensitivity to the ways in which an environment can be inhabited (Lequin 2002) – which is distinguished from that which focuses on local history, or the desire for bucolic natural beauty. It is not a matter of objectifying an enclave to facilitate the experience of curated beautiful landscapes, but rather of welcoming visitors into a living environment. Thus, leading a visit to the forest or the countryside, viewed in this light as a rewilded ruin, touches on the issues of a fast-changing rural world (Bailoni 2012).

5.4. Conclusion: the model of sociology of science and technology as ecological communication

The proposed guidance for visiting wild ruins opens the door to a practice of ecological communication akin to that used in sociology of science and technology: replacing the epistemological critique of world views to assess their relevance in relation to the situations at hand (Latour 2001). Such a visit is not a meditation on the most intrinsically correct way of considering the remnants of the past and of nature, in that it reflects their true nature or is deduced by a method that provides a definite result. The question, rather, is to examine which approach is the most relevant if we wish to relate to the site being visited, taking account of its complex and multifaceted character.

The first step is to recognize that ruins are an environment which cannot be satisfactorily described solely in terms of nature or solely in terms of culture. If we define the experience of visiting as being characterized by that tension, visitors are awakened to the relationship with nature that Bruno Latour (1991) says is characteristic of the modern world. Wild ruins are a place in which modern humans are brought into a relationship with an external environment, lacking the requisite categories in their internal ontology to process it.

Guidance, then, becomes a matter of pragmatism rather than critique. It does not unmask naturalist false knowledge by validating an environmental imaginary whose relevance becomes apparent. To do so would imitate the militant view whereby a site visit is a way of demonstrating to visitors the "correct" approach (Monbiot 2013; Bourg et al. 2021). Far from this prescriptive approach, providing guidance is more about recognizing the dissonance that arises. It is not a matter of creating awareness of a given reality, but of questioning. Thus, the way in which a visit addresses this uncertainty echoes the relational approach formalized in sociology of innovation to organize the airing of opposing views (Callon et al. 2001). It organizes collaboration among stakeholders to deal with what seems an interpretative risk. This approach involves a conception of education which values people's responsibility and commitment, while also endowing them with the skills they need to exercise those responsibilities. By participating, each individual engages in the exercise of their enlightened capacity, in a shared quest to find meaning in their surroundings. The visitors themselves problematize the relevance of their imaginaries, and assess them, taking responsibility for the way in which they engage with the site they are visiting. This requires them to trust in their sensory, reflective and relational capabilities.

This superposition of the communicational dynamics of a visit and the construction of ecological awareness through debate marks the meeting of two types of response to the same situation: second modernity, marked by the loss of belief in expert authority (Beck 2001) and by the lack of an explanatory grand narrative which comes to being accepted through the weight of evidence (Lyotard 1979). Seeking to provide guidance to the participants as effectively as possible, this approach addresses this situation. For instance, it is undoubtedly not the most effective way to produce the shock needed to reshape our relationship with nature, as we so patently need to do, judging by the series of climate disasters unfolding before our eyes (Bourg et al. 2021). If the purpose of ecology is to define incontrovertible solutions which can bring about an acceptable future for society, then we must establish other forms of mediation which adopt a critical approach.

However, even within the scope of its application, the effectiveness of the proposed method depends on a condition, which leaves it greatly weakened. Like ecological policy based on the institutionalization of debate, it depends upon a communicational normative base: in order to be effective, it requires an ideal democratic society in which each member a) is more concerned about fitting in with the others than about cynically imposing their own interests, b) is not irreconcilably ideologically divorced from the others and c) is not in an echo chamber. In the limited context of an organized excursion, people who do not fit these requirements can be excluded. The scope of application, however, is greatly restricted, in a society characterized not only by the lack of authority and of metanarratives, but also by the primacy of individual interests (Fressoz 2012), conflict about the very meaning of politics (Mouffe 2016) and communicational biases (Bronner 2013).

5.5. References

Aït-Touati, F. and Penot-Lacassagne, O. (2022). Mettre en scène la Terre. *Elfe XX-XXI*, 11 [Online]. Available at: http://journals.openedition.org/elfe/4073 [Accessed 16 January 2024].

Bailoni, M. (2012). Introduction : identités, territoires et paysages ruraux en mutations. *Revue géographique de l'Est*, 52(3–4) [Online]. Available at: http://journals.openedition.org/rge/3751 [Accessed 16 January 2024].

Balaën, F. (2002). L'expérience de visite dans les expositions scientifiques et techniques à scénographie d'immersion. PhD Thesis, Université de Bourgogne, Dijon.

Baron, N. and Lajarge, R. (2015). *Les parcs naturels régionaux, des territoires en expériences*. Editions Quæ, Versailles.

Beck, U. (2001). *La société du risque. Sur la voie d'une autre modernité*. Aubier, Paris.

Bensa, A. (2006). *La fin de l'exotisme. Essais d'anthropologie critique*. Anacharsis, Toulouse.

Bigando, E. (2012). L'expérience ordinaire et quotidienne d'un paysage exceptionnel. Habiter un paysage culturel inscrit au patrimoine mondial de l'humanité (Saint-Émilion). In *Patrimoine et désir d'identité*, Fournier, L., Crozat, D., Bernié-Boissard, C., Chastagner, C. (eds). L'Harmattan, Paris, 207–218.

Binnie, J., Edensor, T., Holloway, J., Millington, S., Young, C. (2007). Mundane mobilities, banal travels. *Social and Cultural Geography*, 8(2), 164–174.

Bonneuil, C. and Fressoz, J.-B. (2013). *L'événement Anthropocène. La Terre, l'histoire et nous*. Le Seuil, Paris.

Bourg, D., Demay, C., Favre, B. (2021). *Désobéir pour la terre – défense de l'état de nécessité*. PUF, Paris.

Bronner, G. (2013). *La démocratie des crédules*. PUF, Paris.

Callicott, J.B. (2008). Contemporary criticisms of the received wilderness idea. In *The Wilderness Debate Rages On: Continuing the Great New Wilderness Debate*, Nelson, M.P. and Callicott J.B. (eds). The University of Georgia Press, Athens.

Callon, M., Lascoumes, B., Barthe, Y. (2001). *Agir dans un monde incertain*. Le Seuil, Paris.

Chalvet, M. (2021). Conservation/préservation de la nature "sauvage" : une histoire des savoirs, des imaginaires et des idéologies (XIXe siècle et première partie du XXe siècle). *Revue Française Forestière*, 73(2–3), 227–251.

Charbonnier, P. (2015). *Abondance et liberté*. La Découverte, Paris.

Crosnier-Leconte, M.-L. (1986). *La Gare et l'hôtel d'Orsay (Carnet parcours du Musée d'Orsay)*. Éditions de la Réunion des musées nationaux, Paris.

Davallon, J. (2006). *Le don du patrimoine : une approche communicationnelle de la patrimonialisation*. Hermès-Lavoisier, Paris.

Davallon, J., Gottesdiener, H., Vilatte, J.-C. (2006). À quoi peuvent donc servir les recherches sur les visiteurs. *Culture et Musées*, 8(1), 161–172.

Davin, D. (2020). La visite dont vous êtes le héros. *La Lettre de l'OCIM*, 192 [Online]. Available at: http://journals.openedition.org/ocim/4141 [Accessed 16 January 2024].

De Munck, J. (2023). Les trois dimensions de la sociologie critique. *Sociologie, Research experiments* [Online]. Available at: http://journals.openedition.org/sociologies/3576 [Accessed 16 January 2024].

Debarbieux, B. (1995). Imagination et imaginaire géographiques. In *Encyclopédie de géographie*, Bailly, A., Ferras, R., Pumain, D. (eds). Economica, Paris, 875–88.

Denogent, J. (2022). Notes sur le "primitivisme". *COnTEXTES*, 32 [Online]. Available at: http://journals.openedition.org/contextes/10897 [Accessed 16 January 2024].

Descola, P. (1996). Les cosmologies des Indiens d'Amazonie. *La Recherche*, 292, 62–67.

Descola, P. (2021). *Les formes du visible*. Le Seuil, Paris.

Descola, P. and Pignocchi, A. (2022). *Ethnographies des mondes à venir*. Le Seuil, Paris.

DeSilvey, C. (2017). *Curated Decay: Heritage Beyond Saving*. The University of Minnesota Press, Minneapolis.

Droz, Y. and Miéville-Ott, V. (eds) (2005). *La polyphonie du paysage*. Presses polytechniques et universitaires romandes, Lausanne.

Fressoz, J.-B. (2012). *L'apocalypse joyeuse. Une histoire du risque technologique*. Le Seuil, Paris.

Garrett, B. (2012). Place hacking: Tales of urban exploration. PhD Thesis, University of London, London.

Gellereau, M. (2005). *Les mises en scène de la visite guidée. Communication et médiation*. L'Harmattan, Paris.

Ginzburg, C. (1980). Signes, Traces, Pistes – Racines d'un paradigme de l'indice. *Le Débat*, 6(6), 3–44.

Halbwachs, M. (1994). *Les cadres sociaux de la mémoire*. Albin Michel, Paris.

Halbwachs, M. (1997). *La mémoire collective*. Albin Michel, Paris.

Harrison, R. and Sterling, C. (eds) (2020). *Deterritorializing the Future. Heritage in, of and after the Anthropocene*. Open Humanities Press, London.

Hawken, J. (2016). Philosopher avec les enfants : enquête théorique et expérimentale sur une pratique de l'ouverture d'esprit. PhD Thesis, Université Paris 1, Paris.

Houseman, M. and Severi, C. (2009). *Naven ou le donner à voir. Essai d'interprétation de l'action rituelle*. Éditions de la Maison des sciences de l'Homme, Paris.

Iser, W. (1976). *L'acte de lecture. Théorie de l'effet esthétique*. P. Mardaga, Brussels.

James, W. (2007). *Le pragmatisme*. Flammarion, Paris.

Jonas, H. (1991). *Le principe responsabilité*. Le Cerf, Paris.

Jonk (2022). *Urbex monde*. Arthaud, Paris.

Kohn, E. (2017). *Comment pensent les forêts : vers une anthropologie au-delà de l'humain*. Zones Sensibles, Brussels.

Laoureux, S. (2008). Du pratique au théorique : la sociologie phénoménologique d'Alfred Schütz et la question de la coupure épistémologique. *Bulletin d'analyse phénoménologique*, IV 3, 169–188.

Latour, B. (1991). *Nous n'avons jamais été modernes. Essai d'anthropologie symétrique*. La Découverte, Paris.

Latour, B. (2001). *L'espoir de Pandore. Pour une version réaliste de l'activité scientifique*. La Découverte, Paris.

Le Gallou, A. (2018). From urban exploration to ruin tourism: A geographical analysis of contemporary ruins as new frontiers for urban tourism. *International Journal of Tourism Cities*, 4(2), 245–260.

Lequin, M. (2002). L'écotourisme. Expérience d'une interaction nature-culture. *Téoros*, 21, 3.

Lesslie, R.G. and Taylor, S.G. (1985). The wilderness continuum concept and its implications for Australian Wilderness Preservation Policy. *Biological Conservation*, 32(4), 309–333.

Locquet, A. and Héritier, S. (2021). Interrogation autour de la nature et du sauvage à propos de l'établissement de wilderness areas en Europe. *Cybergeo: European Journal of Geography, Environnement, Nature, Paysage*, 946 [Online]. Available at: http://journals.openedition.org/cybergeo/34986 [Accessed 16 January 2024].

Löwenthal, D. (1998). *The Heritage Crusade and the Spoils of History*. Cambridge University Press, Cambridge.

Luglia, R. (2021). Aux origines des espaces naturels protégés en France. *Dynamiques environnementales*, 47.

Lyotard, J.-F. (1979). *La condition postmoderne. Rapport sur le savoir*. Les Éditions de Minuit, Paris.

Maris, V. (2018). *La part sauvage du monde. Penser la nature dans l'Anthropocène*. Le Seuil, Paris.

Monbiot, G. (2013). *Feral: Searching for Enchantment on the Frontiers of Rewilding*. Penguin Books, London.

Mouffe, C. (2016). *L'illusion du consensus*. Albin Michel, Paris.

Offenstadt, N. (2018). *Le pays disparu. Sur les traces de la RDA*. Stock, Paris.

Offenstadt, N. (2022). *Urbex. Le phénomène de l'exploration urbaine décrypté*. Albin Michel, Paris.

Pereira, I. (2017). *Paulo Freire, pédagogie des opprimés*. Libertalia, Montreuil.

Poulot, D. (1997). *Musée, nation, patrimoine*. Gallimard, Paris.

Poulot, D. (2014). *Patrimoine et musée : l'institution de la culture*. Hachette, Paris.

Roger, A. (1997). *Court traité du paysage*. Gallimard, Paris.

Rosa, H. (2018). *Résonance : une sociologie de la relation au monde*. La Découverte, Paris.

Schiele, B. (2001). *Le musée de sciences. Montée du modèle communicationnel et recomposition du champ muséal*. L'Harmattan, Paris.

Schnitzler, A. and Genot, J.C. (2012). *La France des friches. De la ruralité à la féralité*. Quae, Paris.

Schütz, A. (1998). *Éléments de sociologie phénoménologique*. L'Harmattan, Paris.

Sémal, L. (2019). *Face à l'effondrement. Militer à l'ombre des catastrophes*. PUF, Paris.

Staszak, J.-F. (2006). Voyage et circulation des images : du Tahiti de Loti et Gauguin à celui des voyagistes. *Sociétés et Représentations*, 21(1), 79–99.

Stépanoff, C. (2019). *Voyager dans l'invisible. Techniques chamaniques de l'imagination*. La Découverte, Paris.

Veillon, R. (2021). *Green urbex*. Albin Michel, Paris.

Wadbled, N. (2020). L'imaginaire écologique du tourisme de ruine : faire l'expérience d'une présence de la nature plutôt que de l'histoire. *Téoros, Revue de recherche en tourisme*, 39–2 [Online]. Available at: http://journals.openedition.org/teoros/4547 [Accessed 16 January 2024].

Wadbled, N. (2021). Y a-t-il un patrimoine de l'anthropocène ? Désapprendre ou apprendre à percevoir au-delà de la différence entre nature et culture. *Recherches et éducations*, 23. doi: 10.4000/rechercheseducations.12023 [Accessed 16 January 2024].

Wadbled, N. (2022). L'intérêt environnemental des ruines urbaines : la présence de l'anthropocène au-delà des patrimoines culturels et naturels. *Territoire en mouvement, Revue de géographie et aménagement*, 53–54 [Online]. Available at: http://journals.openedition.org/tem/8645 [Accessed 16 January 2024].

PART 3

Art and Mediation of the Plant World

6

Representation of Plants in Autochthonous Documentary Filmmaking

Since its beginnings, autochthonous documentary filmmaking[1] has taken shape in the image of the overlap between nature and culture in "autochthonous societies" (Descola 2001, p. 14) and has shifted toward a multitude of natures, in line with the statement: "all beings see and 'represent' the world in the same way; what changes is the world that they see" (De Castro 2014), which is inherent to the theory of perspectivism (De Castro 2014). These societies believe that humanity does not stop "at the doorstep of the human species" (Descola 2001, p. 14), in that they are open to "collaboration in their social lives with the merest of plants, and the most insignificant of animals" (ibid.); plants are viewed as having a "similar soul" to humans (De Castro 2014). Films made in these cultures can be taken as a research object (Cyrulnik 2021, p. 3), the aim being to show the intricate relationship between these peoples and the plant world (Brisson 2001, p. 9) as the "complex universe" (ibid.) of their ecosystems (forests, undergrowth, soils, flora including plants, flowers, mosses, etc.). The representation of the plant world in its diversity is addressed explicitly through images or suggested by cultural and traditional representations. It tells us about these peoples' ways of life, in which culture and nature are intrinsically linked. The evolution of autochthonous documentary filmmaking also affords a lens

Chapter written by Marine BRUN-FRANZETTI and Natacha CYRULNIK.

1 "Autochthonous peoples" is the generic term to speak of the earliest people to inhabit the American continent. In Guyana, we speak of Amerindians, but in Canada that term is to be avoided because it refers to the terms used by the colonizers of the Americas. In Guyana, the term continues to be used, but for ethical reasons, in this chapter, we will use the term "autochthonous" or "original community".

through which to view western societies' relationship with the plant world and identify a way in which to think more fully about ecology.

The documentary films made by autochthonous filmmakers (Gergaud 2019, p. 114) chosen for this study should be considered an "ideal cinematographic form", with the aim of committing to "rediscover the truth" and of sharing "their realities" (ibid.). Their ethnicity condemns them to a "status of discrimination victims", leading them to "obtain a cultural sovereignty in the sense of 'authority, autonomy and responsibility in the representation process'". This has echoes of the notion of a "multiverse" (Bona 2021, p. 104) as a way of perceiving the world and nature in a multicultural dimension and as an alternative to a universe – a concept that is inherited from a single and dominant ideology. The multiverse could lead to a form of togetherness, which is crucial to humanity's survival. Thus, filmmakers document the past colonization and the still ongoing confiscation of their land. They are part of the drive to combat the "negation" of the process of decolonization (Lacasse 2015, p. 4) while seeking to convey information about the environmental problems at hand and about their way of dealing with those problems. How do they manage to represent the plant world in sounds and images in their movies? To what end to they do so? To what extent does this representation bring us a fuller knowledge of their situation? As they can reach further than the autochthonous communities, can these representations contribute to environmental defense on a planetary scale? Can an autochthonous filmmaker, then, be considered a primary interlocutor, capable of sparking reflection about the viewer's relationship with the plant world?

With this in mind, we propose to study the cinematographic language of five films, representative of the evolution of documentary filmmaking in today's world. The films have been chosen for the complementarity of their subjects, dealing with the culture of these communities, their history and the way in which they safeguard the environment, in a cinematographic style that is specific to each filmmaker. The question of representation of these territories is at the heart of our analysis, and these five films are representative of it. Thus, we will take these documentary films as objects of study, which will allow us to gain some perspective on the view of the autochthonous peoples and their land, so we know them better. The "re" in "representation" (Bougnoux 2006, p. 53) offers the opportunity to take a step back and gain perspective through our analysis of it. Through these films, we are able to see three phases depending on the given time period in relation to the history of the peoples whom we are seeking to understand. The first phase includes pioneering films made by "mediator–filmmakers" (Biot 2020) (1970 to 1990). The second brings together films made by cinematographers in service of the autochthonous cause (1990 to 2000). The third brings together the work of filmmakers who are committed to the cause of ecology (2000 to the present day). In these three phases, directors film an environment that changes and is constantly degrading over the years. Their way of filming with the aim of putting messages across has changed as audiovisual technologies have progressed. However, the representation of the plant world in its ecosystem is constantly shown as being under threat, and it plays an increasingly important part in autochthonous cinematographic language.

This cinematographic study will enable us to extract data to give us a better understanding of the situation in which autochthonous peoples find themselves. This is precisely what documentary cinema offers, because it allows us to obtain tangible new knowledge about the situation unfolding in front of the camera (Niney 2000, 2002, 2009). Combining cinematographic studies with information and communication sciences, this analysis aims, by the way in which plants are portrayed on film, to decipher the ecological data through image analysis, leading to groups of topics (Paillé and Mucchielli 2012). In the way in which they alter our understanding of people's lived experiences, these movies show the relationship between living beings and their environment, aiming to bring about political and social action, while reminding viewers of the sacred nature of plants in the founding beliefs of these peoples.

6.1. From autochthonous subjects to an autochthonous "filmmaker–mediators", the voice of women to narrate the experience of plants and past colonization

The first phase, spanning 1970 to 1990, is anchored in the geographic and political dimension of the earliest autochthonous-made movies, dealing with the spoliation of the territories and the consequences of colonization – particularly in Canada. In these productions, nature became emblematic of the issue at hand.

The documentary *Mère de tant d'enfants* (*Mother to So Many Children*, Obomsawin 1977, 58') is representative of this first phase on a number of levels. Alanis Obomsawin is a pioneer of autochthonous cinema in Canada and one of the founders of the autochthonous struggle movement. Her films use the cinematographic language of the 1970s documentaries produced by the *Office national du film* (ONF – National Institute of Film), which has produced most of her documentary films. In her view, the voice is "more important than the image" (Mello 2008, p. 37), because that voice is that of an "outside entity, neither more nor less truthful than that of scientific expertise" (Biot 2020, p. 7).

Voiceover is a major cinematic tool that the director uses as the spokesperson for Canada's autochthonous communities. It is didactic and informative, telling the story of the historical situation of the autochthonous peoples to redress the balance of historical justice; the director speaks in the first-person plural: "we". The plant world is part of the narrative in which she speaks of the annexing of autochthonous territories in the form of forests, woodlands and riparian vegetation. Thus, Alanis Obomsawin reappropriates (Remy 2019, p. 16) archive documents from the colonial past, such as a strategy implemented in the context of "political actions and occupations". The film lays down the foundations of the "view" (Scubla 2009) of autochthonous filmmakers in their work, changing from being "viewed" to being the "viewer" (Bastien 2016, p. 39). Her view is simultaneously "distant" (Scubla 2009, p. 5) from the events and places she films, and "situated" within them, because it is directed toward the autochthonous communities she is filming. Her point of view is expressed when she speaks in defense of autochthonous peoples and gives information about the colonization they suffered and the damage done to wild plant life. Through the same mechanism, she tells her audience about her own, and her people's, relationship with the plant world.

Mère de tant d'enfants also features voiceovers from women offering testimonials. This tool is used in a similar fashion to these peoples' strong oral tradition. This tradition refers to a collective dimension that is more "rooted in the past, and in the history of a community, than an individual's own expression. This can be taken as a first distinctive trait in relation to Western oral tradition" (Klaus 1995, p. 122). The point of view of the filmed women is supported by their own voiceover expressed over the image, contrary to previous documentaries in which the narrating voiceover was provided by people outside of the community. Thus, the "traditional" voiceover (Biot 2020, p. 24) is replaced by a "popular voice" specific to Quebec's direct cinema method of the 1960s: "speech acts, coming from below, dethrone the Voice handed down from on high, and establish the reign of veritable collective expression". This "cinema of speech", as it was notably called by Pierre Perrault[2] and Michel Brault[3], who largely contributed to its development, is intended to give voice to everyone, on a level playing field. This reference is important, in that it is the whole of autochthonous filmmaking that is being questioned and examined here. While this form is part of the direct cinema movement, in this context, it can be viewed as a cinematographic genre in its own right, which is emblematic of these increasingly eco-conscious calls. In these films, it is through voice that the plant world is made present, and a form of calling begins to take shape.

This device allows for a coming together of the "filmmaker–mediator" and the "witness" (Biot 2020, p. 24), the person being filmed. Specific to direct cinema, the scenes of the film allow the spectator to be immersed in these women's daily lives. Their role within their family is shown. They trace a portion of their memories and the way in which they live in their family in the forest. The plant world is shown in the form of memories, stated as being part of a life that is no more. For the Ojibwe[4] filmmaker Nancy Morrison, her "religion" is to watch the leaves drop from the trees in fall and walk barefoot in the forest; in her testimonial, she laments that she will never be able to do this again. Thus, these women express their relationship with nature and with their territory, which is in peril. The plant world should be interpreted in its filmed representation by the association of image with voiceover, which supports the filmed image of the natural world. Through these portraits of women and their voices, Alanis Obomsawin constructs a "representation of the world" (Cyrulnik 2021, p. 3). In this context, plants serve as an intergenerational link: the Ojibwe grandmother Marie Williamson and her grandchildren make dolls with leaves collected from their garden. Her granddaughter then balances the leaf dolls in a hammock. Through a voiceover, her grandmother explains that when she was young, she used to play with dolls just like these. Using this sequence, the director shows that plants are used in educating children while also giving them an awareness of the environment. Similarly, plants are put to use by the traditional basket-weaving cooperative, governed by women. Alanis Obomsawin's own voiceover accompanies the presentation of the manual work of weaving dry wicker. Using few tools – a simple pair of

2 Pierre Perrault (1927–1999) was a director and essayist – one of the representatives of Quebec's direct cinema.

3 Michel Brault (1928–2013) was a movie director and director of photography – one of the representatives of Quebec's direct cinema.

4 The Ojibwe are part of the autochthonous community of the Anishinaabe, who live in North America.

shears, their hands and their mouths – these women carry on the tradition and the family heritage. In this context, plants are an economic resource that can be used to make a living. Ancestral know-how is also handed down in the making of plant-based medicinal products. At the school Indian Ways, the pupils learn to handle plants to make the remedies used by their ancestors. The immersive images of pupils with their teacher, and their testimonials to camera, indicate the importance of education in a school run by women and which is dedicated to teaching the languages, cultural practices and history of autochthonous peoples. These schools are shown as safeguarding these traditional activities. The use of voices superimposed on images of plants then indicates a form of engagement, or in any case, immersion, which helps viewers to better understand day-to-day life in those forests. The representation of plants in relation to autochthonous cultures is shown as "an indissociable whole" (Turcotte 2018, p. 19), suggesting an ecological approach, which becomes a source of strength, allowing them to "anchor the autochthonous economic, social and political model".

6.2. Plants as an indicator of environmental degradation: autochthonous struggles and persistence of colonization

The second phase, between the 1990s and 2000s, identifies the change in representation of the plant world in the exploitation of land for agricultural and real estate (for economic purposes), which were consequences of the globalization that took place in these years. It is therefore possible to say that colonization is still in place, but concealed, in the ongoing conflicts prior to effective recognition of these peoples' "right to self-determination" and the right to have their "acquired rights" respected (Russell 2017) – and true liberty and dignity.

The concerns prevalent during this time period are revealed, notably, in the major film, *Kanehsatake: 270 ans de résistance* (Obomsawin 1993, 120'). The filmmaker documents the Oka crisis of 1990, which brought the Mohawks in the community of Kahnawake into conflict with the Quebec authorities. Without a preconstructed scenario, she is "projected" (Saint-Amand 2012, p. 186) into the heart of the events, pulling together the story "in words and images", which are the raw material of her documentary film. The director delivers a narrative that is "articulated in the same way as oral tradition" but without the discourse of an "impartial expert". She affirms the "autochthonous presence" through scenes of resistance and testimonials filmed on the site of the siege involved in defending a sacred forest space – an ancestral burial ground – from being bulldozed to extend a golf course. The Mohawk territory is shown in the form of maps which show the process of spoliation of the lands and the inevitable disappearance of the forests. Alanis Obomsawin's voiceover, accompanying the images, helps to "make plain her own point of view without explicitly stating it" (Saint-Amand 2012, p. 187). Striking and desperate, and sometimes violent, the testimonials from the Mohawks and of the Quebec residents "strengthen" and also "undermine" one another, allowing "viewers to draw their own conclusions on how the series of shots and sequences presented by the documentary fit together". The intentional inclusion of the perspectives of Quebec authorities enables the film to reach a broader audience of Quebec residents. For autochthonous communities engaged in the conflict, *Kanehsatake : 270 ans de résistance* (Obomsawin 1993) is a "podium" from which they can speak, designed to "touch and arouse emotions in the viewer". Through this documentary, Alanis Obomsawin breaks with the institutional documentary style of her early career and displays her desire to put across her own viewpoint as the documentary-maker.

However, while it is these voices that allow for an initial analysis, the place given to nature also helps to underscore the director's own perspective. While the film gives voice to direct witnesses to the crisis – the director herself ultimately among them – the images portray a way of approaching nature that is in keeping with her intentions. The image-to-sound relationship is in keeping with the perspectives adopted by films advocating for the understanding of these territories. In *Kanehsatake: 270 ans de résistance* (Obomsawin 1993), against the backdrop of images of the forest in the Mohawk territory, she discusses the death of Corporal Marcel Lemay, a Sûreté du Québec agent, shot during the conflict. Using handheld camera movements, slow-motion effects, and the voiceover accompanying the narration, the effect created suggests that the corporal's spirit is walking through the trees of the forest. His presence inhabits the forest, suggested by the sunlight pouring through the branches, giving the sense of a spiritual experience. The image of the plants, linked to the image of the corporal when the film talks about his death, is shown as indicative of "the transition to a different life, a different state of being, a different world, and never an end in itself" (Jérôme and Poirier 2018, p. 8). The presence of plants during a homage to a slain man is far from the "idealized vision" (Villalba 2019, p. 2) and "de-essentialized" vision of Nature – a word written with a capital "N". To take off the "capital N" would be akin to breaking down the "ideological barrier" attached to nature, which prevents us from "realizing that everything is interconnected".

While in the 1990 to 2000 period, autochthonous peoples were "considered" peoples fated to die out or as faithful replicas of that imagined figure associated with the ineffective margins of history, politics and the territory (Saint-Amand 2010, p. 84), through her work, Alanis Obomsawin gives them back a place, an existence, at the heart of the modern world. Her commitment, even though the director's own presence within the film is not particularly marked, appears in the way in which she weaves together images and sounds; the film becomes a testimonial in an increasingly ecological register.

We can then wonder about the difference in receipt between autochthonous and allochthonous viewers in connection with their degree of engagement concerning the environment. How can the director make these films into tools for raising awareness? Does she show a specific way of filming as an autochthonous person, capable of exercising influence more widely? How does she allow the allochthonous viewer to better understand so as to engage more firmly in defense of ecology? The filmmakers choose images and sounds to address environmental problems and the specific relationship of their community to nature. They use technological tools as a supplementary vector for the tradition of orality to communicate with a wider audience, combining affect, poetry and imagery.

6.3. A step toward ecological engagement of autochthonous filmmakers

In the third phase, which spans from the 2000s to the present day, the portrayals of plant life in these films have shifted toward the necessity for an ecological transition. While the acceleration of climate change has resulted in crisis situations in autochthonous territories, the documentary films seek to do more than "describe, decry or explain" (Moine 2003, p. 22). One way in which filmmakers seek to emphasize the urgency of the situation, and force viewers to open their eyes to it, is by texts read in voiceover, as "perlocutionary acts" which "aim to change the viewers' behavior, and to produce an effect upon them" (Dupont 2011, p. 22).

The avenues for exploration relate to the influence of the earliest autochthonous films on the next generations of filmmakers – in particular, whether they use the same technical processes as those who have come before them. In these later films, voiceover is used in a more personal and intimate way. The evolution of audiovisual techniques also has influences on the way in which their message is expressed. Their atmosphere (Cyrulnik 2021) helps to affirm the filmmakers' own positions in relation to the environmental problems, which are addressed with the desire for independence and denunciation. They may wish to extend the film's influence beyond their territory by expressing the similarities of globalized environmental problems, which have been widely discussed. Three films illustrate the development of audiovisual techniques in a context where new technologies influence the representation of the plant world.

The documentary film *Unti : les origines* (Yanuwana Pierre 2017, 56') develops the association between the image and the voiceover. Numerous panoramic shots and long takes of the forest along the Haut Maroni in Guyana accompany the footage of travels in a dugout canoe, with the director's voiceover. The inner details of his thoughts are revealed to viewers, expressing his internal narrative as he mourns the loss of his father. The discourse is no longer informative but intimate. The director also decries the loss of the culture of the Kali'na people. His sadness over his father's death resonates with his sadness over the problems of pollution, such as the buildup of waste and gold panning, which are gradually destroying the ecosystem, the flora, fauna and the river. The shots of the Amazonian rainforest taken from the moving canoe make the vegetation a prominent figure within the film, like a character in its own right. In slow-motion shots, sunlight filtering through the branches of the trees indicates the presence of non-humans or spirits. The atmosphere of the film, then, is rendered by the images of these peoples' natural habitat and the soundscape that surrounds them. The zoom in on the forest and the sacred hill of Talwakem focuses the viewer's attention on the luxuriant forest and its expanse. The lively sounds of the animals inhabiting it become louder as the view approaches the hill. The representation of the forest as it is, alive and powerful, is far from the "virgin", "lifeless" and "immaculate" forest (Bona 2021, p. 32) that is typically shown in movies. The forest has a life, and a history, through the animals, plants and spirits that inhabit it. This ecosystem is akin to a representation of nature as something aggressive, sometimes dangerous, as the director implies in his voiceover, numerous times, when he speaks of the forces of nature, the animals and spirits which are in every being. This representation is the polar opposite to an "angelic vision of nature" (ibid.), which echoes the colonial vision of the first explorers, although it could also instill fear when the nature of far-off countries was mentioned. As a member of the Kali'na himself, the director chooses to use the word "we" in reference to his community, with his "individuality" taking a back seat to that cultural identity (Gergaud 2019, p. 125). He presents his own "vision of the Other", participant in the "collective construction of cinema 'about the world'" (ibid., p. 199). His position is as part of the struggle against the French national authorities, who are against the preservation of the environment and the Kali'na community, which is an understandable position in the exercise of their sovereignty over the territory.

Autochthonous narration based on the "art of storytelling" (Dudemaine et al. 2020, p. 4) influences the "form and content of the film narrative, and the creative process". It is a matter of "autochthonization of cinema" rather than "simply the addition of autochthonous filmmakers to the dominant world of western cinema". Autochthonous cinematic language is an act of resistance, presenting alternatives to western representations. The film *Je m'appelle Humain*, by Kim O'Bomsawin (2020, 78'), is an example of that language that brings image and sound together. The presence of vegetation is made into an asset by close-up shots. The voiceover by Joséphine Bacon, poetess and the leading lady in the film, sits over a series of close-ups showing the beauty of the flora (trees, flowers, mosses) and gives an overarching vision of the biodiversity of the land of her birth. These close-ups are linked with Bacon's inner thoughts and her poetry. She directly speaks to the vegetation, in an intimate relationship with nature, like a message to protect it. Thus, both of the films analyzed here are more overtly interested in representing the plant world to spotlight their culture and decry the environmental realities that these peoples are experiencing. Here, the discourse presented in voiceover expresses the importance of the struggle to protect the environment, amplified by wide shots of green life, filmed by a drone. Thus, O'Bomsawin uses contemplative aerial shots to paint a picture of the territory of Bacon's ancestors in the Nutshimit region of Canada. Again, accompanied by her voiceover, the landscape that we are flying over is shown to be scarred by humans, with tractors building a hydraulic dam. The powerful words resonate in the images as an alarm call, decrying the deforestation and destruction of her ancestors' lands. It is a siren call to fight for the engagement of future generations. These images and that voice attest to the colonization of the lands, which is still ongoing.

The use of a drone to support the autochthonous struggle is also found in the film *Ka'a Zar Ukyze Wà* (Guajajara 2019, 13'), where the director goes to meet a previously uncontacted community[5]: the Awa Guaja. The drone is

5 Uncontacted communities are descended from autochthonous peoples. These communities avoid all contact with people outside of their community.

used at the start and at the end of the film, giving a broad view of the environment in which they live. The audience are shown shots of the surrounding forests, the natural habitat of these peoples, reflecting their severance from the world. The filmmaker's voiceover helps the viewer to imagine how the Awa Guaja live.

The use of digital and new technologies allows for a reappropriation of the filmed image of autochthonous peoples, which might correspond to a decolonization of the gaze (Delahaie 2022, p. 1): each filmmaker casts light on themselves in order to "link to their memory" (Spitz 2019, p. 6) and their history and to "all the elsewheres to expand their presence in the world" (ibid.). The use of such tools from the world of western filmmaking serves the message of autochthonous struggle while also allowing the filmmakers to reach larger audiences.

6.4. Conclusion

Our analysis of the representation of plant life in the films studied shows that it is a keystone in the identity of autochthonous peoples, who form a symbiotic relationship with it, and live in harmony with it. This representation is imposed on these peoples as a deep-seated relationship between "humans" and "non-humans". The plant world offers the barometer by which we can measure the harmful consequences of colonization on their way of life and the disastrous damage wrought upon their environment. Thus, it is an important part in raising these peoples' awareness and spurring them to take action. The evolution of colonization can be seen in the continual degradation of plant life over time, despite the autochthonous peoples' struggle against the power of the authorities, whose actions are invariably to the detriment of their ecosystems.

The autochthonous peoples have an interdependent relationship with the plant world and with the other elements of nature. Similarly, filmmakers represent vegetation in order to spark the imagination of the viewer through testimonials, educational or intimate narratives shared in voiceover, the different types of shots used and aerial shots taken by drone. In particular, voiceover used in the films studied here is a truly uniting thread that carries the message being conveyed: "the result of knowledge about the past [is] collective knowledge, shared knowledge" (Vincent 2021). Thus, the words accompany the images of nature to imbue that message with profound meaning and allow the viewer to have a more sensory experience. Its use in connection with the oral tradition produces a sacred dimension, wherein lies the strength of these documentaries. These representations are an invitation to reveal the invisible spirit world, which "governs and enchants the cosmos and everything in it" (Pinçonnat 2023).

From the earliest to the latest films, there is a visible shift in the way in which the filmmakers portray the cultures, rites and practices of these peoples. The most recent tend to "subvert the expectations of a non-autochthonous audience" (Gergaud 2019, p. 193). The more modern filmmakers fully own their desire to "tell a story" that is connected to their own life journey and their struggle. The ecological dimension is increasingly present in the films, thanks to advances in technology such as drones, and also due to the choice of a deliberately sensory approach (Pink 2009), to allow viewers to experience the situation for themselves (Cyrulnik 2015). The director's status as the creator of the documentary film is affirmed: it is more accepted. By analyzing the cinematographic language, we are able to measure the place accorded to nature in these films and in daily life in these places. The directors' action and the people filmed attest to the involvement of humans among the plants. These human interactions take place in a form of "filmic ecology" (Cyrulnik 2018), which is a "genre of film that is particularly open to the Other and to the environment" and also attempts to involve the viewer in that openness. In the autochthonous films studied here, vegetation plays just as important a role as do the humans in the representation of the territory, signposting the way to a different, more ecologically minded, manner of thinking – a different way of seeing nature and living in harmony with it.

6.5. References

Bastien, P. (2016). Filmer l'autochtone. De l'ethnographie à l'autonomisation. *Inter*, 122, 38–42.

Biot, D. (2020). Voix off et voix vive sur fond rouge. Témoignages et images en mouvement dans l'exposition C'est notre histoire : Premières Nations et Inuit du XXIe siècle. *Intermédialités/Intermediality*, 36 [Online]. Available at: https://id.erudit.org/iderudit/1080953ar [Accessed 16 November 2023].

Bona, D.T. (2021). *Sagesse des lianes 1 : cosmopoétique du refuge*. Post-Editions, Rotterdam.

Bougnoux, D. (2006). *La crise de la représentation*. La Découverte, Paris.

Brisson, J. (2001). L'univers végétal des peuples autochtones – comité des végétaux du jardin des Premières-Nations. *Quatre-Temps*, 24(2), 36–37.

Cyrulnik, N. (2015). Le documentaire, un espace de liberté pour une nouvelle communauté. *Revue française des sciences de l'information et de la communication*, 7. doi: 10.4000/rfsic.1744.

Cyrulnik, N. (2018). Le documentaire, vers une écologie filmique. *Registres, Revue des études théâtrales* [Online]. Available at: https://hal.science/hal-01872282.

Cyrulnik, N. (2021). Expérimenter l'atmosphère au cinéma pour construire l'interdisciplinarité. *Revue française des sciences de l'information et de la communication* [Online]. Available at: http://journals.openedition.org/rfsic/10632.

De Castro, E. (2014). Perspectivisme et multinaturalisme en Amérique indigène. *Journal des anthropologues*, 138/139, 161–181. doi: 10.4000/jda.4512 [Accessed 24 June 2024].

Delahaie, F. (2022). Décoloniser le regard écologique, redonner sa place au sensible ? *Les chantiers de la création*, 15 [Online]. Available at: http://journals.openedition.org/lcc/5855 [Accessed 19 November 2023].

Descola, P. (2001). *Anthropologie de la nature. Leçon inaugurale prononcée le jeudi 29 mars 2001*. Collège de France, Paris [Online]. Available at: http://books.openedition.org/cdf/1325 [Accessed 16 November 2023].

Dudemaine, A., Marcoux, G., St-Amand, I. (2020). Cinémas et médias autochtones dans les Amériques. *Revue canadienne d'études cinématographiques*, 29.

Dupont, P. (2011). Étude du documentaire écologiste comme genre cinématographique et de ses impacts individuels et collectifs au sein du mouvement social. PhD Thesis, Université de Montréal, Montreal.

Gergaud, S. (2019). *Rétablir la vérité : le cinéma engagé. Cinéastes autochtones : la souveraineté culturelle en action*. Éditions Warm, Luçon.

Jérôme, L. and Poirier, S. (2018). Conceptions de la mort et rites funéraires dans les mondes autochtones. *Frontières*, 29 [Online]. Available at: https://id.erudit.org/iderudit/1044157ar [Accessed 16 November 2023].

Klaus, P. (1995). Diane Boudreau, Histoire de la littérature amérindienne au Québec : oralité et écriture. *Études littéraires*, 28.

Lacasse, G. (2015). Bruno Cornellier, la "chose indienne" : cinéma et politiques de la représentation autochtone au Québec et au Canada. *Recherches sociographiques*, 58(1) [Online]. Available at: https://id.erudit.org/iderudit/1039956ar [Accessed 1 November 2023].

Mello, M.H. (2008). Entretien avec Alanis Obomsawin/270 ans de résistance. *Ciné-Bulles* [Online]. Available at: https://id.erudit.org/iderudit/33442ac [Accessed 16 November 2023].

Moine, R. (2003). *Les genres du cinéma*. Nathan, Paris.

Niney, F. (2000). *L'épreuve du réel à l'écran. Essai sur le principe de réalité documentaire*, 2nd edition. De Boeck Université, Brussels.

Niney, F. (2002). La poétique documentaire comme forme de connaissance. États généraux du film documentaire. Report [Online]. Available at: http://www.lussasdoc.com/etatsgeneraux/2002/sem_poetique.php4 [Accessed 16 November 2023].

Niney, F. (2009). *Le documentaire et ses faux-semblants*. Klincksieck, Paris.

Paillé, P. and Mucchielli, A. (2012). *L'analyse qualitative en sciences humaines et sociales*. Armand Colin, Paris.

Pinçonnat, C. (2023). Avec la terre en héritage, ou l'éthique environnementale du roman amérindien contemporain. *Revue de littérature comparée*, 386, 237–250.

Pink, S. (2009). *Doing Sensory Ethnography*. London, Sage.

Remy, L. (2019). Décoloniser l'espace. Les documentaires autochtones de l'Office national du film du Canada. In *Colloque Luttes autochtones : formes et pratiques filmiques*. PSL Research University, La Fémis.

Russell, C. (2017). Le colonialisme canadien, d'hier à aujourd'hui. *Nouveaux Cahiers du socialisme*, 105.

Saint-Amand, I. (2010). Retour sur la crise d'Oka : l'histoire derrière les barricades. *Liberté*, 51.

Saint-Amand, I. (2012). La crise d'Oka lors du siège, dans les films documentaires et dans les récits littéraires autochtones et allochtones au Québec et au Canada : événement, rapport à l'espace et représentations. PhD Thesis, Université du Québec, Montreal, 186–187.

Scubla, L. (2009). À propos du regard éloigné de l'ethnographe ou les rapports entre la théorie et l'observation. *Ateliers du LESC*, 33 [Online]. Available at: https://journals.openedition.org/ateliers/8199 [Accessed 16 November 2023].

Spitz, C. (2019). Essai. Pour émerger un cinéma autochtone dans le Pacifique. *Journal de la Société des Océanistes*, 148, 127–132 [Online]. Available at: http://journals.openedition.org/jso/10145 [Accessed 15 November 2023].

Turcotte, Y. (2018). Our way of life should be seen as a natural economy : le rapport à la nature et l'identité autochtone dans The Fourth World de George Manuel. *Cahiers d'histoire*, 36 [Online]. Available at: https://id.erudit.org/iderudit/1054184ar [Accessed 16 November 2023].

Villalba, B. (2019). Timothy Morton, La Pensée écologique. *Développement durable et territoires*, 10 [Online]. Available at: http://journals.openedition.org/developpementdurable/16458 [Accessed 1 November 2023].

Vincent, S. (2021). La tradition orale : une autre façon de concevoir le passé. Les Autochtones et le Québec. Des premiers contacts au Plan Nord. PhD Thesis, Les Presses de l'Université de Montréal, Montreal, 75–91.

Filmography

Guajajara, F. (2019). *Ka'a Zar Ukyze Wà*, 13 min.

Obomsawin, A. (1977). *Mère de tant d'enfants*, 58 min.

Obomsawin, A. (1993). *Kanehsatake 270 ans de résistance*, 120 min.

O'Bomsawin, K. (2020). *Je m'appelle Humain*, 78 min.

Yanuwana Pierre, C. (2017). *Unt+ : les origines*, 56 min.

7

Art to Convey Biodiversity: Analysis of an Augmented-reality Animated Sculpture

This chapter presents the results of a joint research project[1] ongoing since November 2021, relating to a sculpture enriched with an augmented reality app , entitled *Immersion, Nature Augmentée*[2]. This scientific mediation device pertaining to biodiversity represents a cross-sectional slice of a river bank and aims to raise public awareness about the need to protect wetlands. It was commissioned by the Bouches-du-Rhône local authority from the Conservatoire d'espaces naturels (CEN – Conservatory of the Natural Environment) for the Provence-Alpes-Côte d'Azur (PACA) region. The project was cofunded by the water board Agence de l'eau Rhône Méditerranée Corse. This sculpture was born of the Bouches-du-Rhône local authority's desire to have an immersive piece to help educate various types of audience about how wetlands work and how their soils are structured. This desire came in response to a notable lacuna in scientific mediation, because such spaces rarely feature in mediation. This corresponds to a concern of the local authority for Bouches-du-Rhône, which includes the Camargue (which also occupies territory in Gard); this region is considered

Chapter written by Marie-Caroline HEÏD, Sarah LABELLE, Emma LAURENT, Valérie MÉLIANI and Eva SANDRI.

For a color version of all figures in this chapter, see www.iste.co.uk/souillard/emerging.zip.

1 The research group is made up of all the co-authors of this chapter and is led by Valérie Méliani.

2 See a presentation of the sculpture on the website of the studio, *Les Fées Spéciales*: les-fees-speciales.coop/portfolio_page/immersion-la-nature-augmentee [Accessed 12 December 2023].

one of France's largest wetlands, with its lagoons, ponds and freshwater swamps. Also, wetlands are shrinking natural environments: the surface of wetlands in France shrank by 50% between 1960 and 1990. This observation can be linked with the ecological crisis, which numerous scientists, including the IPCC, are warning of.

Our research explores this complex work of art, which serves to give media exposure to a natural environment that is undervalued. More specifically, the problem that sparked this study can be summed up by the following question: how can we understand the tensions between art, nature and technology in the mediation offered by the sculpture *Immersion, Nature Augmentée*? We study these tensions at various levels: from the issues of design by the project owners, the artist and the professional animators who created the augmented sculpture to the activity of reception by audiences and the affordances (Gibson 1977) of the artwork and its augmented reality system. We will begin by examining the relationship between art and science, the issues at stake and the limitations in scientific mediation. We then present the sculpture *Immersion, Nature Augmentée*, before going on to detail our research framework. Finally, we discuss the results, in turn analyzing the three forms of the mediation on biodiversity, which overlap when the audience receives the information: firstly, artistic mediation on the part of the sculpture; then technological mediation, which is provided by the augmented reality system and finally, the discourse of the mediators about the sculpture.

7.1. State of the art: relationship to the living world and sensitivity crisis

How is the combination of art with science designed to contribute to changing the face of scientific mediation about nature? In what way do the artistic representations of physical phenomena and living beings offer a better understanding of the environmental issues of today's world? Our study shows how the modes of scientific communication address the contemporary issues of the environmental crisis and how the choices of forms and mediation objects play a part in helping users to grasp and understand the environmental knowledge that the sculpture itself conveys.

7.1.1. *Using mediation objects to render the knowledge concrete*

Research on the diversity of scientific mediation has shown the complexity of the processes (Jeanneret 1994) and the multiplicity and creativity of forms of representation. While the field has long been dominated by a model of vulgarization whereby scientific texts that are felt to be unreadable to non-experts are rewritten in plain language, the relevance of that model has been questioned since the 1980s; in addition, we observe a rich figurative system that is a crucial part of scientific popularization (Jacobi 1985). Indeed, the practices of scientific mediation are anchored in this idea of the "figurability" of problems and knowledge, helping to illustrate them and make them more easily understood. Daniel Jacobi points out that in vulgarization texts, "imaging" plays a two-fold role: reference to the scientific domain, in the form of images taken from actual scientific research and an educational function, with illustrative images that help to explain the science and help it to stick in the recipients' minds. In addition to this are the sometimes humorous drawings, which anchor the discourse in an area other than that of science. Thus, we see the co-presence and articulation of forms from the scientific domain, serving to imbue a text with authority, with forms that play with the reader's sensory imagination. Presenting things in a figurative manner is a way of helping the audience to see and to organize their perception of the knowledge. These reflections about the different types of mediation objects raise questions about the possible combinations to offer a sensory experience to audiences thinking about science and open the door to the analysis of systems such as the sculpture *Immersion, Nature Augmentée*.

Studies on vulgarization have demonstrated links between the choice of representation and the modes of scientific communication: these links offer a way of connecting with the knowledge. All of these tools – analogies, metadiscourse, metaphors, etc. – are designed to facilitate the sharing of knowledge. They are part of a logic whereby science is put out into the wider world and involve multiple levels: the status and accessibility of the knowledge or scientific debates and controversies. These studies highlight how the choices made in terms of mediation objects have impacts on the possible relations to the knowledge presented and represented. From this perspective, Camille Jutant and Julien Bogroff studied a pedagogical project in which student designers were asked to create a representation of quantum physics, without the aim of making it accessible to an audience of non-scientists (Jutant and Bogroff 2015). The benefit of this work, which can be viewed in relation to what we sought to observe with the sculpture in our study, is that it sheds light on the communicational aspect of the design and the different "meaning-creation processes" in the systems created. One of the main results highlights the plasticity of the forms used in a design project and the variety of relations to the knowledge objects that this allows and encourages: participants took everyday objects and identifiable forms and built them into a narrative or construct relating to quantum physics. The absence of any constraints in the production of mediation objects in design is also seen in art: the exploration of forms allows for a freer relationship with the knowledge at hand, Camille Jutant and Julien Bogroff explain. Thus, their study indicates the way in which reflexivity manifests in the project design and is based on a fairly broad degree of freedom offered by the nature of the relationships between designers, project owners and mediators. In our study, we again see this variety of actors engaged in the process of production of composite mediation objects. As we will see, this configuration has implications for the constraints and freedoms for each player as they create representations of wetlands and the associated knowledge.

7.1.2. *Seeing the living world through the mediation of art*

This research sits at the intersection between the two axes of the "processes of production of environmental communication" and "media and [...] cultural representations of the environment", according to the typology defined by Hansen and Cox (as referenced by Catellani et al. 2019).

In our study, the mediation project is based on the representation of a wetland; therefore, it is sensible to focus on how that living environment is made visible, and how the audiences are brought into contact with the knowledge. The choice of art is in line with a tradition of representation of the living world: art shines a light on nature, from cave paintings to landscape paintings to still life paintings. However, these modes of representation contribute to a way of understanding the living world and defining the relationship that humans have with nature. In a way, art illuminates the relations that human beings have with nature and the living world and with non-humans more broadly. Philippe Descola, through his anthropological research, demonstrated the dichotomy between nature and culture in western societies, based on the formalization of the notion of culture (Descola 2005). This distance is notably manifested in the artistic forms that have fed our imaginaries and dominated our representations of nature (Zhong Mengual 2021). Estelle Zhong Mengual, following in the footstep of other art historians, points out the existence of a deficit of imaginaries in relation to the living world and how certain arts have helped shape the way in which we see the living world – i.e. have taught us how to see and not to see simultaneously.

On the basis of a study of landscape paintings, Baptiste Morizot and Estelle Zhong Mengual (2018) explained that our relationship with the living world and natural environments is structured by the dichotomy between art and science. In the view of these researchers, the ecological crisis we currently face stems from the way in which we establish and maintain a relationship with the living world and the view of it which has been constructed over the ages through art. The ecological crisis must be thought of in connection to a "sensitivity crisis", meaning difficulties in developing "forms of attention and qualities of availability" towards the living world. These difficulties are attributable to the way in which landscape painting has shaped our view of the living world by creating distance from the natural environment. This culture is partly responsible for urban and western humans' inability to connect with the environment. Numerous art historians, cited by Morizot and Zhong Mengual, point to the fact that landscape paintings establish a relationship with the environment that is dominated by perspective and depth – i.e. a broad view in which there is no place for the viewing subject, who is placed in a position of exteriority. This visual culture is prolonged and maintained through photographic practice. We can stress the fact that conventional painting leads to the formation of a particular viewpoint, giving primacy to the sense of sight and not affording the opportunity to explore other senses such as touch and smell. Our cultural history teaches us to regard an object from afar and to contemplate it. In modern art, however, we see the emergence of artistic devices that make use of other senses and play with them, notably through immersive experiences.

When faced with the observation of a formatted regard, what we need to do is relearn to produce interpretation processes, given that the culture of landscape painting has taken us away from understanding the living world. It therefore becomes necessary to reshape the practices of interpreting the world in order to better observe our surroundings (Jeanneret 1994; Haraway 2009). How can knowledge of the relationship with nature and with the living world be aligned with knowledge of mediation objects and processes? How do the practices of mediation of natural environments contribute to bringing humans into a closer relationship with the environment? Our study analyzes the way in which the sculpture *Immersion, Nature Augmentée*, is part of that re-mediation of the interpretation of the environment and the relationship with nature.

7.2. Presentation of sculpture

7.2.1. *Origins of the project*

The sculpture was born of an original collaboration between multiple partners who were known to each other before engaging in the creation of the system. The presence of an early version of the sculpture at the 2021 congress of the IUCN[3] (International Union for the Conservation of Nature) in Marseille helped CEN PACA to garner support from financial backers.

Therefore, the Bouches-du-Rhône authority wished to donate to CEN PACA a new version of the sculpture – smaller and mobile – which could be taken to different places and events: educational establishments, media libraries and various events surrounding biodiversity. The original work, exhibited at the IUCN congress, is referred to as the "monumental sculpture", and the smaller version is known as the "voyager module".[4]

a) *b)*

Figure 7.1. *The sculpture Immersion, Nature Augmentée. a) Monumental sculpture and b) voyager module (source: Les Fées Spéciales)*

3 "The French Committee of the IUCN currently brings together two ministries, seven public institutions, five local authorities and 57 non-governmental organizations, as well as a network of more than 250 experts gathered within thematic commissions and working groups". See: https://uicn.fr/qui-sommes-nous/ [Accessed 5 December 2023].

4 Monumental sculpture: width 3.7 m, depth 1.8 m and height 2.2 m. Voyager module: width 1 m, depth 0.70 m and height 1.10 m.

Artist Anne-Lise Koehler[5] was commissioned to produce the sculpture, designing it with cloth and papers taken from books, rescued from bric-a-brac stores, the Emmaüs Foundation, or destined for the dump. The "recycling of resources and the absence of any toxic materials"[6] was a choice made by the artist, adopting an eco-friendly approach, which reflects the issues of respect for biodiversity that the project itself spotlights. The design of the tablet-based augmented-reality system was handled by the animation studio Les Fées Spéciales.

This studio develops systems for cultural and scientific institutions in the area of sustainable development and the environment. The sculpture was designed in close collaboration with the scientists of CEN PACA, who ensured that the information conveyed by the system was accurate. All of these actors (the artist, the animation professionals and the scientists) worked together for around a year to create both versions of the sculpture, defined as an "immersive mediation tool combining art, nature and technology".[7]

7.2.2. *Presentation of the sculpture as a mediation object*

Immersion, Nature Augmentée offers viewers two types of mediation on the subject of wetlands. Initially, without any assistive tools, visitors are invited to walk around the sculpture to discover its content in a naturalist and artistic way. The structure is designed to be examined one face at a time, season by season, beginning with the face representing springtime. Depending on the context in which it is exhibited, the public can view the sculpture from a few meters away or from up close: there is no form of limitation, either symbolic or physical, in the space around the sculpture.

5 A writer and director of animated films, scriptwriter, painter and sculptor, she has notably developed a naturalist body of work with sculptures made of paper. She has collaborated with Les Fées Spéciales on many occasions: for the graphic direction of the sets and characters for animated films including *Kirikou et la Sorcière*, co-headship of special effects and co-direction of documentaries such as *Bonjour le Monde !*

6 Statement taken from a communication about the *Immersion, Nature Augmentée* project.

7 Statement found on CEN PACA's website. See: https://cen-paca.org/participer/expositions-itinerantes/immersion-nature-augmentee [Accessed 2 December 2023].

However, although there is no notice to say so, viewers are not allowed to touch the sculpture, and the mediator will intervene if anybody comes too close to the artwork to prevent any contact. In this way, the mediation device is endowed with artistic value as an untouchable object, to be viewed only, because of its fragility. From very close up, however, it is possible to read certain passages from the books from which the elements of the sculpture were made, such as the words "*rire*" (laugh), "*rêver*" (dream) and a passage from *Pantagruel*.

Figure 7.2. *Immersion, Nature Augmentée: details on the sculpture*

Secondly, using a digital tablet that works with augmented reality, audiences access an enriched informational universe. The sculpture provides an additional documentary space using the augmented-reality digital tool (Tardy and Dobedei 2015; Sandri 2020). In this documentary space, on the screen, users can touch certain areas of the sculpture represented.

Figure 7.3. *The sculpture Immersion, Nature Augmentée, representation in augmented reality animations*

On the app, there are two available options: a contemplative and free experience and a knowledge-based visit. The latter offers three modes: “strolling – water cycle”, “understanding wetlands” and “discovering flora and fauna”. We can interpret these modes as access to types of knowledge categorized by theme, with some categories being more complex and denser than others.

Once they have selected their desired mode, the audiences are guided to look at all four faces of the sculpture for the elements that can be animated or can activate textual content. If they want to, they can change mode by going back to the menu. In the informational journey, the interface gives users access to an augmented-reality animation for a period of 12 minutes on average. On the screen, against the image of the sculpture itself, they can see new animals and insects (fish in the water and tiny turtles coming out of the ground). Above the sculpture, we can see elements of weather such as clouds or rain. To access scientific information, the viewers have to target elements of the sculpture with a yellow dotted target that is constantly present at the center of the interface.

Figure 7.4. *The sculpture Immersion, Nature Augmentée, access to scientific data*

7.3. Studying a hybrid mediation system: fields and methodology

As this sculpture is designed to be a mediation device, below, we describe how to study the peculiarities of this hybrid system, which combines artistic, technological and scientific aspects.

7.3.1. *The space in which the sculpture is exhibited: the different fields of research*

Our study focuses more on the voyager module (which we were able to follow to exhibitions in different spaces) than the monumental sculpture, which is not similarly designed to be easily moveable.

Created so as to be transportable, the voyager module was on show throughout 2021 at the Écomusée (Ecomuseum) in La Crau and was often exhibited in different spaces and at various events (farmers' conferences, national biodiversity conference, science fairs, etc.).[8] We studied the module

8 The list of places and events where the voyager module has been exhibited is available on the CEN website. See: https://cen-paca.org/participer/expositions-itinerantes/immersion-nature-augmentee/.

on four sites. In November 2021, we occupied the first exhibition space over the four days of the CEN's annual conference in Tours, where the sculpture was first exhibited to expert audiences (professionals, mediators and researchers). We then followed the device to the Saint-Chamas media library for two days in February 2022 to present it to school-age audiences (aged 7 to 10). We continued to observe it on March 22, 2022 at the Sainte-Catherine-de-Sienne school in Aix-en-Provence, which hosted *Immersion, Nature Augmentée* for the day to present it to the classes for 6-year-olds and 7- to 8-year-olds. Then, we were able to supplement our observations when the module was exhibited at the Camargue and Rhône Delta Festival, in Port-Saint-Louis-du-Rhône, in late May 2022.

Remaining the property of the artist, the monumental sculpture is difficult to transport. Notably, it was shown among other digital devices, for the exhibition of the ICONIC Festival, organized at Halle Tropisme in Montpellier in November 2022. This space, in a cultural third place, during an event for both the general public and professionals around the cultural and creative industries (CCIs), is the last event that we observed, in July 2024.

7.3.2. *Methodological framework*

In each of these fields, we followed a similar methodological protocol comprising in-depth observations and semi-directed interviews of verbalization in the course of action (Vermersch 2019). Note that the diversity of our fields provided us with data about diverse audiences (schoolchildren, professionals in scientific mediation and CCIs, pupils' parents and families). In the context of our observations, we were attentive to different elements in three major phases: discovery of the work, exploration of the finer details of the sculpture and, finally, the use of the tablet to discover the animated sculpture. We approached these five fields as a team, generally made up of three members of our research group. That approach allowed one of us to concentrate on observing the discourse of the person in charge of the mediation, while another conducted a floating observation of the situation (analysis of how the space was managed, how the audiences moved around the sculpture, interactions among visitors, etc.), and the last member focused on the audiences' interactions with the sculpture and with their peers. In parallel, we conducted short interviews during and after they had experienced the composite system to discover the audiences' feelings and opinions.

In-depth, comprehensive, qualitative interviews (Kaufmann 1996) were also conducted with the coordinators and producers of the project, members of Les Fées Spéciales, scientists and the CEN PACA mediator. Finally, we took advantage of the space to analyze the sculpture as a sociotechnical device (Peeters and Charlier 1999). Thus, we tried to determine all the ways in which it was possible to make use of the sculpture – or, in other words, its affordances (Gibson 1977). Primarily, we looked at the tablet-based animation system; we analyzed the texts, the way in which they were presented, and the user instructions, in order to better understand the appropriation conditions that the designers built into the system.

7.4. Tensions between the artistic, technological and scientific dimensions: the main research results

In this section, we examine the relationship between the artistic, scientific and technological dimensions in the matters of design, the affordances of the system and reception by audiences. From this perspective, it is important to specify that *Immersion, Nature Augmentée* comprises three levels of mediation:

– The sculpture in itself serves the artistic mediation of an alluvial forest. This first form of mediation takes place at a visual level, which brings art and nature together.

– The augmented reality tablet serves for the technological mediation of the sculpture. This second layer of mediation is somewhere between the artistic, the scientific and the sensory.

– Finally, the mediators' discourse adds a third layer of mediation relating to biodiversity. In the final case, the discourse is essentially scientific and tends to lose the artistic dimension.

Below, we describe each of these levels of mediation and their implications in terms of design, affordances and reception.

7.4.1. *The sculpture: the artistic mediation of nature from a sensory point of view*

The sculpture is based on various imaginaries of mediation, stemming from connected and closely associated devices.

7.4.1.1. *Imaginary of the scientific mediation system*

The use of day-to-day materials (cardboard, paper and glue) in the sculpture is reminiscent of certain scientific mediation tools (such as pedagogical toolkits), which can be found in ecomuseums, natural parks and scientific and industrial culture centers (CCSTIs, from the French). Also, it should be pointed out that both children and adults tend to try to touch the sculpture, probably thinking that it is another mediation device. The sculpture, despite its recognized beauty, is not always endowed with the aura of being a work of art, which would cause visitors to suppress their desire to touch it (spaces in Aix-en-Provence and Tours). It seems that the established conventions of consideration of works of art at a distance in an institutional context such as that of a museum are not carried by the sculptural device, particularly when it is exhibited in the context of scientific mediation (the space in Tours). Rather, that context encourages normative practices which are found in the discovery or experimentation by means of handling and use of the senses – notably, the sense of touch. This device and the spaces in which it is exhibited include "not only the body in representation, but also the way in which the body is experienced by the act of receiving" (Martin-Juchat 2008). Thus, we can question the utilitarian reception of this sculpture, shaped by the context in which it is exhibited, by also looking at its format. Composed of different recycled materials, worked by hand, it deconsecrates the distancing required when we are viewing artworks, instead encouraging the audience to get close to the sculpture, both physically and symbolically, through the interactions produced by handling the tablet and walking around.

7.4.1.2. *Imaginary of the artwork produced through recycling*

In the same sense, the artist's choice to take an eco-friendly stance (using books donated to Emmaüs or destined for landfill as the primary material for the sculpture) reveals a link between the recycled material and the message of raising awareness inherent in the work, aiming to encourage environmental protection. The discourse accompanying the sculpture also specifies the imaginaries sparked by this material: "Paper is a very human material. It carries our culture, encoded in symbols and dreams in the natural and wild world" (linguistic elements pertaining to the structure, CEN PACA working document).

7.4.1.3. *Imaginary of the figurative and naturalist device*

Finally, the figurative aspect of *Immersion, Nature Augmentée* is also reminiscent of the artistic universe of naturalist sculptures of animals – notably the work of Mélanie Bourlon, an ultrarealistic contemporary artist. Currently, numerous practices of scientific mediation and exhibitions are based on logics whose aim is to represent wildlife and show the invisible.[9] Indeed, when they are discovering the details of the work, many audiences highlight its virtuosity, which is due to the realism of the sculpture. They exclaim: "that's so well done!", "that's mad!", "you'd think they were real!" (Aix-en-Provence) and try to touch the flowers or animals. The mediation professionals stressed the extent to which the scientific accuracy goes hand in hand with the figurative aspect. The sculpture, as a figurative representation, a "clone" of the wetlands, is designed to correspond to the scientific reality and to have "evocative power" (Jacobi 1985). The naturalist goal is in tension with the artistic form and support in the sculpture and resonates with the desire to bring a poorly known natural environment to light.

It should also be stressed that the structure, presented as a cross-section of the wetland environment, provides an even more unprecedented dimension, in that it shows a part of reality that is not visible in day-to-day life: the inside of the soil. The possibility of discovering a reality that usually escapes us is one of the reasons for the amazement audiences feel when they see the sculpture.

7.4.2. *The augmented reality system: a technological mediation of the sculpture*

Secondly, after approaching the work without the help of any tools, and without being able to touch it, audiences are invited to take up the tablet and access an enriched world. The river and its bank come alive through the animations, and a variety of texts appear, presenting information about the biodiversity of wetlands and the water cycle. Thus, the augmented reality brings a new artistic dimension, which is at the intersection between mediation of the work and scientific mediation.

9 For example, see the exhibition *Mini-Monstres*, which was run at the *Muséum d'histoire naturelle* in 2022, focusing on acarians, microscopic parasites and other tiny fauna.

7.4.2.1. *Tensions between imaginaries concerning nature and scientific data*

These augmented reality animations give rise to a second stage of amazement. The audiences are caught between two opposing injunctions: the figurative representations of nature and the scientific data on biodiversity.

7.4.2.1.1. An artistic expression in the service of the scientific data

Having collaborated many times on past projects, Les Fées Spéciales and the artist are recognized for their practice and have developed an affinity in the expression of their creativity. This experience has given them a shared vocabulary which can be used to bring together scientific data and artistic form. These two worlds are designed for complementarity, or continuity, with one referring to the other, and vice versa.

7.4.2.1.2. Phenomena with the magic touch

School-age audiences (primarily children aged 6 to 7 and 7 to 8) exhibited highly enthusiastic reactions to the augmented reality animations. They were impressed, and they expressed their surprise when the dynamic content showed animals moving. Just like the children, adult audiences were also struck by seeing and hearing the rain falling on the vegetation to differentiate the seasons. Whatever their age, they were captivated: “That’s magical!”, “It’s magic!” (Aix-en-Provence) and drew attention to the accompanying elements: “You see the rain?”, “That’s mad! It’s raining!” (Montpellier). The emotions stirred up by the sculpture, and amplified by the animations, were ultimately used to draw audiences in, to direct their attention to the knowledge about biodiversity being conveyed. The head of CEN PACA describes that dual objective and states that “the emotion in itself would almost be enough to raise awareness”.[10] As Gaëlle Crenn and Jean-Christophe Vilatte (2020) indicate, emotions play a key role in the experience of a visit, “firstly because the visit evokes emotions in the visitors – emotions which the designers then seek to actively evoke and steer; and secondly because emotions are integral parts of the experience of mediation between the visitor and the exhibition, however that mediation is presented”. Also, the beauty of the sculpture and the surprise carried by the animations allow the visitors to get an impression of the environment that is being conveyed, before that impression is supplemented by the scientific mediation. The sculpture *Immersion* offers a clear illustration of situations in which

10 Interview conducted in November 2021 in Tours with the head of the CEN PACA Bouches-du-Rhône chapter.

"emotion can improve the learning or the experience: in that sense, we can say that it is a re-mediation" (ibid).

7.4.2.1.3. Proxemics and social distancing from the artwork

When a visitor gets too close to the sculpture, the tablet displays a message asking them to step away. This message could be interpreted as being reminiscent of the rules in museums whereby audiences are asked to keep a set formal distance from the works exhibited in institutional spaces. Thus, the sculpture is considered fully in its artistic dimension, protected by boundaries that are not to be crossed; otherwise, the visitor will lose access to the animations and scientific data.

7.4.2.2. Tensions between mediation of the artwork and scientific mediation

The members of the studio Les Fées Spéciales have indicated that, in addition to its artistic purpose, the animation also has the twofold objective of "guiding the visitor in discovering the work, whilst respecting the scientific data".

These various tensions, which occur between the mediation of the artwork and the scientific mediation, manifest themselves at various levels – notably during the visitors' journey of discovery of the sculpture. The visit begins with the face representing springtime and should logically continue with the summertime face, which is on the left-hand side of the sculpture. We noticed, in our observations, that audiences did not know which direction to go around the sculpture. Some of them proceeded anticlockwise and therefore moved from the springtime face to the winter one. The weather conditions are supposed to guide users, because they can orientate themselves in relation to the passing of the seasons. We found, however, that the audiences did not necessarily understand this. Firstly, in order to see the weather elements, they have to point the tablet upwards and step back to give themselves the space to make larger movements. Certain users, therefore, did not notice this information. Secondly, the animations correspond to the real weather conditions in the seasons observed. Rain is an element that is common to fall and spring. This scientific reality does not conform to the collective imaginary, as audiences tend to think of spring as being a sunny season

7.4.2.3. *Tensions between the esthetic aspect and the scientific mediation*

Remember that the device has three modes, which offer different types of information. We observed that the mediator only launched the third mode. When we met with the designers, they explained that the first two modes were too complex.

We continue our analysis with a study of the mode most commonly used by the mediator: flora and fauna. Firstly, as was the case for circulation around the sculpture, we find that there was no set path or hierarchical organization of the information that could guide the users. Then, in order to access the information, the audiences were required to use a target at the center of the screen to view points linked to the sculpture (red for animals and green for plants). This interface takes them into the video game universe. The fact of having to target points in order to access content is reminiscent of the gameplay in the smartphone game *Pokémon Go*, which involves finding and then capturing imaginary creatures. Once the points have been targeted, boxes appear, offering either specific information about a plant or animal species, or more general information about the water cycle or wetlands. In the former case, the data are fairly superficial and are displayed in a small box in the lower center of the screen. In the latter case, the text takes up the whole of the interface. Overall, we found that the audiences did not take the time to read the information – particularly in the case of the longer texts. In interviews conducted after the visits, schoolchildren were unable to recall the information on biodiversity that they had seen on the tablet. Thus, the correct balance between art, technology and science does not appear to have been struck. The esthetic attractiveness of the animation and the "magic" effect of the augmented reality on the audiences, coupled with the videogame-like imaginaries, led them to focus on finding possible ways to interact with the exhibit, skipping from one to the next without actually concentrating on the scientific data.

7.4.3. *Discourse constructed about* Immersion, Nature Augmentée*: human mediation about biodiversity*

When we observe the audiences' interactions, it becomes clear that human mediation is necessary, in view of the technical aspects of the system. The app, and therefore the augmented-reality part of the experience, launches only after scanning a QR code on the front of the sculpture, which means that a mediator needs to be present whenever the tablet is used. Based on the areas on which we have focused, we have been able to identify two scenarios: firstly, discourse delivered by the scientific mediator recruited by CEN PACA to accompany the mediation of the voyager module, and discourse produced by a cultural mediator about the monumental sculpture.

With respect to the voyager module, we were able to reconstruct the discourse that the CEN PACA mediator was specifically trained to deliver during the course of three field investigations at a media library, a school and a public festival on the Camargue. From her first interactions with the audience, the mediator presented the sculpture not as a work of art, nor indeed as an educational resource, but as a game, the aim of which is to find how many animals there are. To groups of schoolchildren viewing the sculpture, she introduces her discourse by saying: "there are loads of animals hidden in here; it's your job to find them all!" This challenge immerses young audiences in a game-like universe, which continues to unfold as they use the tablet to discover the animations. In addition, the mediation discourse does not approach the sculpture as a work of art but as a pretext to mention the elements of scientific knowledge linked to wetlands. The arrangement of the sculpture is used as a starting point for discussion with the audiences. Such is the case, for example, when the mediator tells the students: "spiders eat mosquitoes, and we thank them for that!" (Aix-en-Provence). This utilitarian dimension of the art to serve science is observed in a western-type relationship with the living world (Descola 2005).

The second mediation situation that we observed was the event at the Halle Tropisme, which was designed both for members of the public and for professionals in CCIs. Here, a mediator attached to the festival itself was detached to guide audiences in interacting with the monumental sculpture. Due to its central position, its monumental form and its bright colors, the sculpture stood out from the other devices on display, which were very digital-focused. The mediator immediately took visitors in hand, directly offering them the opportunity to discover more about the sculpture and the environment that had inspired it, using the tablet. Thus, few of them had the opportunity to view the sculpture in its own right first, without the tablet. Unlike in the contexts described above, the visitors spoke of the feeling that the system was designed primarily for young audiences. Finally, we observe that the context of the exhibition strongly influences the audiences' experience of the sculpture, which they view, alternately, as a system designed for children, for adults or for mediation professionals.

7.5. Conclusion

During our observations and interviews, the audiences seemed to be primarily touched by the sculpture's esthetic beauty and by the technical brilliance. The sculpture allows audiences to access sensory knowledge about wetlands, which is encouraged by a somewhat enchanted relationship with nature, established by the whole system. The technological mediation gives the illusion of having a physical relationship with the natural world by holding the tablet in your hands and moving your body (up and down and moving around) to trigger the mediatized interactions with the sculpture. Thus, this system fits in with contemporary practices of scientific mediation, as described by Fabienne Martin-Juchat: "All cultures have established rules that are designed to contain expressions of affect; thankfully, they have also developed the technical dimension, artistic, industrial, manufactured artifacts, etc., because these offer ways in which we can express our emotions" (Martin-Juchat 2020).

Unlike landscape painting, in which the viewer is excluded from the scene, being only an outside observer, the system discussed here invites audiences in to browse the ecosystem of wetlands without many constraints. This offers the possibility of closer contact with the natural world and with the observation: the aim, plainly, is to encourage audiences to pay attention to the natural world through movement and by clearly linking climate phenomena and biodiversity. Specifically, there is the suggestion of learning to observe by moving your eyes: the system helps create pleasure in observing.

7.6. References

Catellani, A., Pascual Espuny, C., Malibabo Lavu, P., Jalenques Vigouroux, B. (2019). Les recherches en communication environnementale. État des lieux. *Communication*, 36(2) [Online]. Available at: http://hdl.handle.net/2078.1/219175 [Accessed 14 December 2023].

Crenn, G. and Vilatte, J.-C. (2020). L'émotion dans les expositions. Introduction. *Culture et Musées*, 36, 15–33.

Descola, P. (2005). *Par-delà nature et culture*. Gallimard, Paris.

Gibson, J. (1977). The theory of affordances. In *Perceiving, Acting, and Knowing*, Shaw, R. and Brandford, J. (eds). Lawrence Erlbaum Associates, New York, 67–82.

Haraway, D. (2009). *Des singes, des cyborgs et des femmes. La réinvention de la nature*. Éditions Jacqueline Chambon, Arles.

Jacobi, D. (1985). Sémiotique du discours de vulgarisation scientifique. *Semen*, 2 [Online]. Available at: http://journals.openedition.org/semen/4291.

Jeanneret, Y. (1994). *Écrire la Science. Formes et enjeux de la vulgarisation*. PUF, Paris.

Jutant, C. and Bogroff, J. (2015). Objets de médiation de la science et objets de design. Le cas du projet "Design quantique". *Communication et Langages*, 183, 9–24.

Kaufmann, J.-C. (1996). *L'entretien compréhensif*. Nathan, Paris.

Martin-Juchat, F. (2008). Penser le corps affectif comme un média. *Corps*, 4, 85–92. doi: 10.3917/corp.004.0085.

Martin-Juchat, F. (2020). *L'aventure du corps. La communication corporelle, une voie vers l'émancipation*. Presses universitaires de Grenoble, Grenoble. doi: 10.3917/pug.marti.2020.02.

Morizot, B. and Zhong Mengual, E. (2018). L'illisibilité du paysage. Enquête sur la crise écologique comme crise de la sensibilité. *Nouvelle revue d'esthétique*, 22(2), 87–96.

Peeters, H. and Charlier, P. (1999). Contributions à une théorie du dispositif. *Hermès, La Revue*, 25(3), 15–23.

Sandri, E. (2020). *Les imaginaires numériques au musée : débat sur l'injonction à l'innovation*. MkF, Paris.

Tardy, C. and Dodebei, V. (eds) (2015). *Mémoire et nouveaux patrimoines*. OpenEdition Press, Marseille.

Vermersch, P. (2019). Entretien d'explicitation. In *Vocabulaire des histoires de vie et de la recherche biographique*, Delory-Momberger, C. (ed.). Érès, Toulouse, 340–342.

Zhong Mengual, E. (2021). *Apprendre à voir. Le point de vue du vivant*. Actes Sud, Arles.

PART 4

Sensory Aspects and Narratives Surrounding Human–Nonhuman Relations

8

A Few Lessons to Draw from Ayahuasca About the Knowledge of Togetherness

8.1. Introduction

Ayahuasca is a beverage made from a mixture of a shrub and a vine, which has psychoactive effects. This chapter offers a reflection from an ethnographic study in Bolivia and Peru, which led to an encounter with this plant, used by *curanderos* (traditional healers in Latin America), and which has become popular in global culture portrayed in the media. Examining what that encounter might mean, this work follows in the footsteps of that of Abram (2010), touching on the ethical, intimate and political consequences of sensory interrelations with the more-than-human world (in a qualitative sense) – that world of relationships that goes beyond what our cultures include and forces us to adopt an attitude of humility when we give it true consideration.

In the context of the ontological turning point in social sciences and in anthropology in particular, it is no longer possible to content ourselves with studying the diversity of representations we hold about the real world. What we need to do is seriously consider the ontological diversity of perspectives about the properties of existing beings (Mol 2002) and thus reshape the social sciences to include nonhuman entities. These evolutions consist of a shift in methodology that enables us to think about relationships with non-humans in terms of interdependence, rather than in terms of duality. A collection of researchers has worked on integrating the question of the living world into our approaches in social sciences, starting with the work of

---

Chapter written by Philippe HERT.

Haraway (2016). For these scholars, the living world is defined as openness to alterity, which places us and our knowledge in a position of vulnerability (Lalande and Le Marec 2022), not in the sense of a "loss", but in the sense of necessary openness to what has not been previously taken into account.

Hence, they suggest that we need to move past the dichotomy between naturalist and animist perspectives and focus on the agency of non-humans (Despret 2013), meaning the assemblages and associations between humans and non-humans that give rise to new capabilities for action and new stances. This allows us, finally, to look at possible modes of alliances: we are in a period of adjustment, exploration of relations with existing beings that we have previously neglected (Morizot 2023), and finally considering them to be living beings with which we need to interact and coexist with. The important role in social sciences that has recently been taken by this invitation to form alliances with non-humans, and the revisited notion of a living being, elucidates the experience of which we are speaking here.

8.2. The field study

The starting point for this field study, some results of which have previously been published (Hert 2012), was the World People's Conference on Climate Change, in 2010, chaired by Evo Morales in Cochabamba (Bolivia), which gave rise to the Universal Declaration on the Rights of Mother Earth. Meeting representatives of autochthonous populations on that occasion, I was led to continue the study for a month, in an isolated Amazonian village with a population of Tacana, focusing on the research question of the material limitation, the boundaries of modernity's influence on peripheral spaces, spaces with limited circulation of western objects and technologies and proximity in spite of distance and (non-)meeting between worlds. Upon returning to La Paz, I met an anthropologist working on the evolution of shamanism and, in particular, ayahuasca as an example of the mix between tradition and globalization. The subject had come up when she told me of a Shipibo shaman and his family in Peru, who was developing a practice with Westerners but without spilling over into a lucrative activity for tourists and without the practice requiring belief in spiritual or religious interpretations (Dupuis 2022). I decided to go and meet the shaman, for a further three weeks, in Peru. The idea was to experience a different limit – that of a possible encounter with the radical alterity of plants. For this purpose, I adopted an experiential ethnographic approach (Goulet 2011) in order to gain sensory knowledge of what could happen in this context.

The aim was to include, in this approach, the possible sensory interrelations with the more-than-human world. I then supplemented this exploration in 2019 by collecting eight testimonials from members of an association in Peru that offers cultural stays in a traditional center where "ayahuasca healing" is practiced.

8.2.1. *How can we make a connection with plants?*

In order to speak of a relationship with plants, we will focus on a concrete experience – i.e. an embodied experience. According to Varela's concept of enaction, the body is the essential element of cognition, beyond a sterile opposition between knowledge (external and objective) and affect (internal and subjective), or between objective knowledge and experience. Thus, the field study is fundamentally linked to the researcher's physical presence and that of the other people whom they meet during their study. Paying attention to what is felt when people are in direct contact is often disdained because it is viewed as being tainted with emotion, according to an approach in sciences where field studies are viewed only as a way to capitalize on data in the market of academic productions, rather than an arena for conversation, availability and multiple forms of attention (Le Marec and Faury 2023). Due to the pretext of rejecting a local, partial and affective perspective, the relationship to knowledge in sciences remains non-incarnate – i.e. nowhere. However, this overlooks the fact that any attempt to gain knowledge and objectivity is shaped by stances which are anchored in material, social and historical context. Taking account of the anchored nature of knowledge allows us to finally break free of the major division between objectivity and subjectivity, between subject and object, between rationality and belief and between body and mind. This echoes the perspective of standpoint epistemology (Harding 1992), which does not refer simply to the idea of a multitude of "standpoints", but to an awareness, constructed through time and through collective action, of being caught up in the interaction of dynamic forces. It is always a partial standpoint, whether it is dominant or dominated.

8.2.2. *Contexts in which the plant is used*

The social, economic and ideological context of our relationship to the plant generally constructs an initial general framework through which we view the experience. Ayahuasca has had an important historic role in Amazonia, helping to build bridges between Amerindian worlds, and toward a non-Indian, urban world, even becoming part of Peru's national cultural heritage (Lozonscy and Mesturini Cappo 2011, p. 221). Rituals connected to the plant can be detached from their original cosmological context and form part of a logic of negotiations and shifting representational arrangements that bring together the vehicles of the idealized image of "indigenous" with people seeking ayahuasca, in an exchange of spiritual and therapeutic resources for improvement of material conditions: "thus, ayahuasca is the turning point […] because it continues to be represented and interpreted as an entity endowed with agentivity – therapeutic, legal, political, ecological and patrimonial" (Lozonscy and Mesturini Cappo 2011, p. 225). Westerners' powerful attraction to plant-based shamanism, the quest for "authentic" experiences and the exoticism which they carry have developed into a tourist industry, notably in Iquitos in Peru, publicized on social media. Amselle (2013) decries this model: he speaks of neocolonialist shamanic tourism and industrial exploitation of the plant; we could add extractivism to this list, because of the capitalist reappropriation of traditional practices, which are radically transformed by the very fact of being exploited for capital gain.

8.2.3. *The agentivity of plants*

Whilst industrial exploitation is real and problematic, it is because there is a great deal of interest in the numerous forms of agentivity that can be constructed with plants on an individual level. Amselle's macroscopic analysis does not take this into account. The very experience of the plant entails considering what it does to us, rather than only what we do to it. Here, we are in agreement with Myers who, through her "affective ecology" theory, shows how knowledge in plant sciences has an incarnate, sensory and affective dimension, which is not seen in the scientific literature but which can create experiential and existential shifts (starting with Darwin, who was struck by the forms of reciprocal arrangements between orchidaceae and bumblebees and demonstrations of sexual attraction) (Myers et al. 2020).

In parallel, an important point is that we must ourselves avoid doing that which we seem to condemn others for doing, i.e. practicing human sciences in an extractivist way, by collecting experiences linked to traditional knowledge and then reusing them to generate knowledge that is detached from the context in which they were produced. The work presented here is by no means intended to encourage readers to take such a neo-shamanic trip to Amazonia, or to play down the dangers of drug use for mental health, but sets out to elucidate certain aspects of the attractiveness of experimenting with this plant and, in particular, the promise of an *encounter* with it. In my view, this term refers to an experience of displacement, which I have documented from a first-person perspective, in full awareness and at a distance from the aspect of modernity that tends to use alterity and integrate it into its own logic, without making the effort to think differently about it.

To begin with, we will look at the relationship with the plant through the lens of the system's material approach: the *set and setting* (Dupuis 2021) – i.e. the contextual dimension of the expectations that influence the experience, what the person has in mind at the time, and also the location, interactions, music or chanting that provide structure to the experience. Thus, we will look at the systems and their importance. We refer to Servais and Halloy's (2013) analysis of the experience of enchantment from the perspective of the anthropology of communication. Their analytical framework appears to be a helpful starting point. We then extend the analysis, using a perspective fed by science studies, to take account of our ability to make connections, even beyond the human world.

8.3. The three stages of the experience

The framework presented by Servais and Halloy (2013) notably relates to encounters between humans and animals, which they describe as magical encounters. We will examine the appropriateness of this description of enchantment, which they see as being based upon a mental construct of interpretations and expectations relating to the lived experience and a material construct that corresponds to a framework of reference points. Certain parallels can be drawn with the experience of an encounter with plants; these will be highlighted below. We will then explore an avenue that goes beyond the effect of enchantment and of the expected promise (an entirely Western idea) of surprise and a "powerful and authentic" experience.

Servais and Halloy (2013) distinguish three steps in their process: (1) The time prior to the encounter, which is already part of the system. This first step is that of imaginative conditioning and education of attention; (2) the encounter itself and what the individual learns from the experience, lived in the first person, and no longer through the perspective of others or observation, enabling them to create experiences of taking the substance (Bessy and Chateauraynaud 2010); (3) the social evaluation of the experience which was lived and the modes of engagement which were brought to bear

(Houseman 2002) – i.e. the ways of connecting to the surrounding world (human and nonhuman), which includes a coherent set of expectations, perceptions, emotions and specific actions.

When we apply this in the context of an encounter with the plant, the first step corresponding to delving into the abundant body of literature. The people who wish to have an experience with the plant generally come with some sort of prior baggage: non-conventional medicine, psychotherapy, spiritual approaches or bodily practices. In preparing for the trip (Lombardi 2021), so-called esoteric bookshops and YouTube videos may be a useful first stop, in that, through books, testimonials, objects, talks and workshops, within western societies, they explain "traditions" in connection with "educational plants". The success of these bookshops is testament to the desire to have an encounter, but at the same time, they provide a stable and continuous discourse about the "authentic" relationship with nature enjoyed by early peoples, without ever taking into account the real contexts of practices linked to ayahuasca. Dupuis (2022) indicates three types of reasons in the Western world to go looking for an encounter with the plant: therapy, spirituality and the search for a political alternative to reconnect with nature and counter-culture. There are also various types of spaces in which the use of the plant is developing: forests, urban environments in connection with Santo Daime – a religious movement that began in Brazil in the 20th century that uses ayahuasca as a sacrament – and finally, laboratories, where there is growing interest from the pharmaceutical industry in the plant's effects – notably in psychiatry and in treating addictions (Sanabria 2021).

8.3.1. *Narratives in the field*

When I entered the field for this study, I had not previously read the literature on ayahuasca; I arrived there by an unconventional route, without being familiar with Western discourse surrounding the plant. I was interested in how a shaman recognized by their peers worked with Westerners who had stepped away from the recreational consumption easily available in urban environments, to join them in their space so far removed from anything else. In a traditional context, when I met the *vegetalista* (traditional healer, using plants; herbalist), he asked me during a short conversation in Spanish what I had come to seek and then proposed a protocol for how to proceed: a strict diet (banana or rice), slowing down and isolating myself from the world for several days:

> It's now my third day in this hut, where there are absolutely no possible distractions, and I really don't feel like reading the only novel that I've brought with me. I am still wondering whether I did the right thing by setting out on this experience. Right now, I'm less sure. This is all so long. So very, very long. Maybe a transformation has already begun to take place, with the diet and the isolation in the forest. I am listening to every sound, animals, my senses are developing, the smells, paying attention to the light, all the insects that pass and see me, sometimes indifferent to my presence. At the same time, though, I am here – this is what I was searching for: the edge of civilization (Field diary).

Then came the first time I took ayahuasca, at night. I had a conversation with the shaman the next day and then took the substance a second time two days later. The context in which we take it is crucially important and has as much impact as the molecule itself. Firstly, visitors are asked to spend a period before the experience slowing down, listening, engaging in introspection and paying attention to the world around them. The plant causes violent vomiting before or after the hallucinations set in (if they do). For certain *vegetalistas*, this is the essential wellbeing effect (bodily purging) that the plant produces, the visions sometimes being only secondary (although that is not what Westerners come in search of). Finally, a third essential aspect is the singing and whistled melodies while you take the drug. In a traditional context, only the *curandero* (healer) takes the plant (Westerners have asked to do so), and the ritual melodies that they intone after taking it are supposed to cure the patient. According to numerous testimonials which I have since gathered, the plant brings about a state of presence of tranquility, attention to the effects of purging, to the singing, to the natural environment around the taker and leads to forms of permeability, with the outside world and with the world within, with the subconscious, and also a blurring of the lines between the internal and external.

Everything happens as though the plant were not acting directly as medications do but creating a context. The context, the set and setting influence the experience, i.e. the dimension of expectations, what we have in our minds at the time, and also the place, the interactions, the music or singing and will alter the structure of the experience. However, here, I wish to highlight the dimension of encounter rather than that of the system.

Then came the third stage: that of social evaluation of the lived experience and writing about it. Among the testimonials, some speak of a veritable experience of encounter through an organic experience:

> I felt myself completely relieved and carried away […] the plant was telling me to come, and I felt my body being transformed in order to understand what ayahuasca wanted to tell me and to show me.

The experience involves a reassessment of our own place, relationships and conceptions of the living world. There is astonishment from some, who wonder how a plant can so radically change the way in which they view the world, because of the attention to their environment that the experience provokes, and a form of marveling, and the fact of being "guided" in this way by a nonhuman entity outside of themselves. Numerous online testimonials speak of entering into a relationship with an "intelligent" nonhuman entity:

> I truly understood why it was spoken of as a teaching plant […] it made me close my eyes, and I started talking to someone, but I knew that I wasn't hallucinating […] it talked to me […] I heard that voice, so lucid, clear, and sonorous.[1]

> Mother Ayahuasca took me in her arms, told me I was her beloved child and gave me a love I never could imagine would exist […] it was really the love of a goddess […] I know she doesn't lie […] it's not about drugs, I took drugs, I know it's totally different, this is not an illusion, Ayahuasca was holding me in her arms for two hours […] I could see her, I could touch her and she could touch me.[2]

For Westerners, the experience produces a need to produce meaning, put it into words and explain the meaning of the visions, statements, of that time spent outside of time and ordinary life:

> After my first, disquieting experience of feeling physical transformations throughout my body and a physical displacement to other spaces, last night's experience gave me a lesson in humility. I wanted to ask questions of the plant, ask it

1 See: https://www.youtube.com/watch?v=jfeEHx9bg1U.

2 See: https://www.youtube.com/watch?v=NxucMOfZEkk.

> to tell me what it knows, teach me things. And there, there was nothing. It was only when I followed the melody of the singing that I got back the "visions of presences". I understood that I am always too inclined to project what I want to see, to know, what I had in my head and absolutely wanted to confirm, instead of simply listening, seeing, letting it come, remaining patient, and accepting that I may not necessarily understand (Field diary).

There is indeed a significant gap between the pragmatic practices of *curanderos* like "rural doctors" and Westerners' spiritual expectations or demands for psychological help when they come to consume the plant. The object here is not to weigh in on the debate surrounding the setting of practices relating to the use of the plant, but to shed light on the way in which perceptions connect to users' expectations, and how the modes of engagement allow us to tie them to their experiences, perceptions and emotions. A comparison with the magical encounters that Servais talks about would narrow down the object of my remarks.

8.4. Dolphins and plants

Véronique Servais speaks of magical encounters with dolphins (marketed as "dolphin experiences"), as amazing experiences of direct communication with the animal, which are felt as revelations. These are experiences which go beyond a simple encounter with an animal, because there is an exchange that takes place, "as though the dolphin was talking about myself with immense and love and benevolence". She analyzes these experiences through the lens of a social system of learning that feeling of enchantment. It is based on an environment (the ocean) which is capable of producing sensory saturation, an intense emotional experience and a bodily experience out of the ordinary, notably referring to the techniques of ecstasy studied in anthropology. Those who experience it speak of a coming together in the most powerful sense, an interspecies encounter, from one conscious subject to another. What makes this possible is an education in attention for the diver, an arrangement of perceptual saliences in an unfamiliar environment and a social evaluation of the perceived experience:

> Enchantment takes place in the context of an "intermediary space" whose cognitive, emotional, perceptive and material elements are all held together by a system. That system creates a link between a set of expectations and an extraordinary

> experience of communication with an animal, by creating the conditions for such communication to take place. The system creates a promise of surprise in the sense that the uninitiated lives the experience "in body and soul" on the basis of what they already know, and is shaken to the core by it (Servais and Halloy 2013, p. 275).

> Seeing the ocean so immense, […] the swimmer feels extremely small, vulnerable and helpless, at the mercy of the ways. Only the boat connects him to the human world. Then, the dolphins swim up. They are big, powerful and agile, […] and, unlike the human, are in their element (ibid., p. 282).

There are various parallels that can be drawn with that which, according to the authors, produces the experience of enchantment: the material framework that is constructed, which allows the individual to be immersed in a sensory environment, with the experience coming in a context that is both rich and indeterminate: an intense emotional experience linked to a bodily experience out of the ordinary, the participant's body and mind are stimulated by the situation, the promise of surprise, the search for settings which give us the elements for interpretation of our lived experience, a focus on mental states (Luhrmann et al. 2010) and the discovery of the agentivity of plants.

This analysis is based on the stance of an observer who is outside of the encounter themselves. The descriptive language it offers focuses on the context and the expectations in connection with the sensory experience and on the discursive dimension of the experience. While that dimension is important in order to socialize the experience, as we have seen through the above testimonials, the content of the experience and what it gives rise to in those who experience it have yet to be defined. The system helps to create an intermediary space and the settings to allow an imaginary to be deployed and then to be updated on the basis of the experience (Belin 2002). In addition, in the dolphin experience, the animals give clear signs of interaction, and social affordances, as Servais terms it (matching the swimmer's speed through the water, looking them in the eyes, inclining their heads, etc.), which a plant cannot do. However, if we look deeper, beyond the system of enchantment and discursivity which is employed, there are arrangements with the plant which become perceptible in their own right. For this purpose, it is necessary to change our descriptive language and focus more on agentivity in a situation that relates to the self, in order to better reflect the nature of the encounter.

8.5. An encounter with the plant

The experience itself is rapidly taken up in all sorts of interpretative frameworks, and there is a saturation of discourse (for example, mystic, therapeutic and political): discourse focusing on the consequences, issues for them, for the planet, for our relationship with the world, as well as mystical revelations and healing. A whole body of literature has developed, which looks at the experience in the expected interpretative frameworks (the third phase described by Servais). This makes it more difficult to give credence to the testimonials, taken in these pre-existing frameworks, and means we need to base our thinking on the sensations and thoughts in the situation rather than on the interpretations. With a detached analysis of the experience, it may be difficult to home in on what is at stake. The first-person experience allows us to better understand what makes sense for the people to whom we have spoken, and clarify how, for some people, it is a transformative experience. Of course, the question arises of how to describe the experience in the social sciences, how to write about it, the legitimacy of the first-person experience, and the bodily and mental involvement of the researcher in the field – in short, of knowledge which is grounded in an individual's lived experience.

8.5.1. *The search for radical alterity*

If we look only at the descriptions of encounters with the plant, they highlight the fact of interacting with a conscious, living "entity": "you can't cheat with the plant", "the plant was teaching me", "I was in physical contact with entities" and "I received a great deal of love". Participants interact with the plant within their own bodies. Quite literally, the plant is within them (or rather, a mixture of two plants is), and its molecules are binding to theirs, causing them to see images, feel their bodies and sometimes hear sounds, which are not the direct effect of interaction with their environment. Unlike medical usage of next-generation psychiatric drugs (Nichols et al. 2017) or recreational drug use (though the emetic effect of the plant is off-putting), here the testimonials highlight the encounter that is made tangible through an intermediate state in which we cannot tell whether our perceptions come from ourselves or from the plant while having the feeling of remaining fully conscious. As with any encounter, what happens is in the space in between, in the exchange, or in the adjustments that we make in order to interpret the message from the other being (is the plant adjusting to us? In the context of an animistic interpretation, we could say that). While it is a matter of semiotics here (Kohn 2017), it is also a question of the capacity to experience a relationship with alterity.

In relation to dolphins, Servais tells us that what has the effect of producing enchantment is the expectations, the context, the system itself and the perceptive saliences. Here, this enchantment may also take place for the same reasons, but the inner experience, when it happens, is surprising in a different way to the encounter with an animal. The surprise is that of the discovery of an alterity acting within ourselves, or rather, which redefines what the self is. This search for interaction and encounter, for spiritual, therapeutic and political reasons, or the quest for knowledge, leads to a connection with an agentive nonhuman alterity "of nature". This link causes us to experience a shift, both in our perception and in our bodies, and can also be experienced as receiving care. It is an intimate relationship, "from one conscious being to another", with the plant. In the experience, this intimacy is already created by the fact of having ingested the plant: it is literally part of us. More than that, though, what we encounter is the agentivity of the plant: it makes us think, it gives us bodily experiences which shake our perceptions and we are no longer sure of who we are. There is an assemblage (Despret 2013) which blurs the lines between us in ourselves and that entity which allows us to exist differently. Then, we only have discursivity to try to give meaning to the experience and validate it from a social perspective.

The aim here is not to establish whether or not the plant "truly" makes us see or "truly" speaks to us, but to take, from that experience, the ability to have an encounter with a radical form of alterity, even though it is based on an imperfect understanding (Servais and Servais 2009). Indeed, every experience with an alterity, including a plant, feeds our capacity to feel, to get involved and to make meaning on the basis of what that relationship does to us. David Abram's fundamental position is that we need to (re)consider how we exist in the world, which "opens up the idea that every natural element has the potential to be expressive … 'everything speaks to my very flesh as a creature', and it is at this fundamental level of expressivity that the origin of our human language must be situated. […] [which echoes] the interpretation that we can make of Merleau-Ponty's notion of flesh – a concept in which every being is part of an expressive relationship with all others" (Kristensen 2021, p. 151). The quest for an encounter with the plant is an experience of possible and unprecedented modes of alliance. It is a way (explicit, for certain people) of repoliticizing our experiences of encounters with non-humans, in the sense of collectively re-inhabiting the connections between humans and non-humans, even though they contain a certain dose of imagination.

8.6. Epistemic issues: taking account of care

While it is important to take account of the set and setting, as a framework that conditions the experience, it is also important to consider the effects of displacement caused or produced by the lived experience, such as a slowing down (which is required by the context in which the plant is traditionally taken), increased attention to the space, to others and to ourselves, through the ritual forms in which we are invited to take part (which are as many forms of engagement and ways of paying attention) and acceptance of what we do not know. All of that points us to the ethics of care – i.e. an effort to pay attention, constantly and locally, without a generalizing and abstract principle, to maintain a safe framework and pay attention to relationships (Mol et al. 2010), even with the plant. It may be that these efforts and deliberate attention, due to the willingness to have an encounter with the plant, which are transformative, rather than simply the context in which the plant is taken. In addition, the pharmaceutical industry looking closely at the plant falls down where the context of paying attention to the situation (whether traditional or otherwise) becomes highly important: "[Ayahuasca] is not generally understood to work as a drug (in the magic bullet sense of the word) but rather to facilitate or initiate a process of healing" (Sanabria 2021, p. 395).

These experiences touch on our ability to feel, to be surprised, our capability for relation and involvement and our capacity to change our viewpoint, including for a researcher looking into these practices. These qualities are, a priori, not those attributable to a scientific approach, unless we consider a perspective grounded in knowledge and place relationships at the heart of the investigative approach (Le Marec and Faury 2023). Thus, the researcher's involvement comes into question. To have the experience of a relationship, it is not possible to retain this (illusory) stance as a neutral scientist, detached from the object of their study, as though they were unaffected by it. The knowledge-based perspective, on the contrary, causes us to shift our focus to the way in which we relate to and communicate with our "objects", enabling us to produce new descriptions.

8.6.1. *Re-establishing a connection with plants*

This is precisely the subject of Gagliano's (2018) book, *Thus Spoke the Plant*. The author, in an approach that she describes as antidisciplinary, establishes a parallel of equal importance between her shamanic experience with plants which led her to change her research area from marine biology to plant biology, and her major discoveries in plant bioacoustics (plants perceive and produce sounds). She calls for an overall shift, and cautions against anthropocentrism, whereby we would humanize plants. On the contrary, it is because we humans are moving into a different space that communication with the plant can occur:

> And for thousands of years, indigenous healers, witchdoctors, and shamans around the world have been learning the songs of plants as a way of communicating with these other-than-human persons and acknowledging them as the guarantors of human existence, the true philanthropists of the world. A relationship built on this premise can only engender a deep sense of gratitude and humbleness that awakens our role as custodians with duty of care and respect toward these other beings and the vital knowledge they enrich with us (Gagliano 2018, p. 24).

These stances illustrate the degree to which detachment from an anthropocentric way of thinking allows us to see how, in reality, plants have given us life. Myers speaks of the Planthroposcene, because it is plants which made this world inhabitable for us animals, who breathe oxygen ("we are of the plants") and they continue to sustain us (Myers 2018). Why could we not enter into a relationship with them (material, symbolic and even mental), if our own existence is the result of theirs? If the plant's molecules have such an effect on us when we ingest them, does that not speak to our shared roots (certainly far back) and thus, a portion of existence that we can share together?

These experiences of ontological diversity, which seem to be undergone in the body and not in the mind, mean that the body, and the body of the researcher, becomes an arena for experience. The plant acts by transforming the relationship, not just symbolically or cognitively, but also by producing a bodily, emotional, cultural and also spiritual reaction, as has often been attested by those who have experienced it. It is not simply a matter of representations but of an experience of gratitude and humility.

8.7. Conclusion

At the start of this chapter, we asked what it could mean to have an encounter with a plant. Beyond the experience of enchantment which is produced and which occurs, we saw that it could give rise to a fleshly experience that renders tangible an ontological shift, and which opens up perspectives on knowledge of the relationship and the shift. The following are three such shifts:

– Envisaging a move away from anthropocentrism, which is not an easy thing to achieve. We cannot break free of our own ontology; that would be illusive. It is not a matter of pretending to be animistic. On the contrary, these experiences of an encounter with alterity, where it is not easy to actually experience it in the ordinary world (from this point of view, an encounter with an animal is easier to envisage), commit us to a practice of humility, acceptance of not knowing and paying attention to that alterity, from the perspective of care. These practices enable us to cultivate a different kind of perspective on knowledge based on relationships and on the researcher's commitment.

– Envisaging the return of enchantment in our relationship with knowledge. This involves making room for living beings and for emotions. Doing so will allow social sciences (HSS) to produce other narratives, free of our colonial perspective (on knowledge, on objects to be collected, on their patentability, on experiences to be lived, on the visions of others, on their ontological assemblages), to finally consider how other ontological experiences, which are more than just "pure subjectivity", can alter the way in which we see the world.

– Finally, that agentivity, which is achieved with the plant, leads us to pay attention to all of the ways in which we are connected to non-humans and opens us up to a new approach to these relations which goes beyond the division between natural sciences and HSS. It also allows us to look at what connects us and drives us to involve ourselves in the lives of all beings (Myers 2018), rather than remain in a position where our only relationship with non-humans is one of utility, exploitation and control.

The final word in this narrative on plants goes to an autochthonous writer:

I am the grandfather yagé. I am the spirit who stays on his feet.

I am healing.

The god who gets you drunk on wondrous dreams.

How many illnesses have I blown away with my breath?

My word is of life and of knowledge[3].

8.8. References

Abram, D. (2010). *Becoming Animal. An Earthly Cosmology*. Pantheon Books, New York.

Amselle, J.-L. (2013). *Psychotropiques. La fièvre de l'ayahuasca en forêt amazonienne*. Albin Michel, Paris.

Belin, E. (2002). *Une sociologie des espaces potentiels. Logique dispositive et expérience ordinaire*. De Boeck, Brussels.

Bessy, C. and Chateauraynaud, C. (2010). Retour sur "Le savoir-prendre". *Techniques et Culture*, 54–55.

Despret, V. (2013). From secret agents to interagency. *History and Theory*, 52(4) 29–44.

Dupuis, D. (2018). Apprendre à voir l'invisible. Pédagogie visionnaire et socialisation des hallucinations dans un centre chamanique d'Amazonie péruvienne. *Cahiers d'anthropologie sociale*, 17, 20–42.

Dupuis, D. (2021). Psychedelics as tools for belief transmission: Set, setting, suggestibility and persuasion in the ritual use of hallucinogens. *Frontiers in Psychology: Personality and Social Psychology*, 12. doi: 10.3389/fpsyg.2021.730031.

Dupuis, D. (2022). The psychedelic ritual as a technique of the self. Identity reconfiguration and narrative reframing in the therapeutic efficacy of ayahuasca, Hau. *Journal of Ethnographic Theory*, 12(1), 198–216.

Gagliano, M. (2018). *Thus Spoke the Plant*. North Atlantic Books.

3 "Kue unaodikue, Joriaina naidaidikue; Jira namadikue; Jifaiya buinaimadikue; Digaa duiko duiñode funotimiedikue; Kuemo jiyua baitara uai ite" (translated by Arthur Malcolm Dixon from Candre's Spanish version of her poem).

Goulet, J.-G. (2011). Trois manières d'être sur le terrain : une brève histoire des conceptions de l'intersubjectivité. *Anthropologie et Sociétés*, 35(3), 107–125.

Haraway, D. (2016). *Staying with the Trouble: Making Kin in the Chthulucene*. Duke University Press, Durham.

Harding, S. (1992). Rethinking standpoint epistemology: What is "strong objectivity". *The Centennial Review*, 36(3), 437–470.

Hert, P. (2012). Pour une prise en compte du corps sensible dans la recherche de terrain : un savoir communicationnel. *Questions de communication*, 22(2), 251–272.

Houseman, M. (2002). Qu'est-ce qu'un rituel ? L'autre. *Cliniques, cultures et sociétés*, 3(3), 533–538.

Kohn, E. (2017). *Comment pensent les forêts*. Zones Sensibles, Paris [Original published 2013].

Kristensen, S. (2021). Les randonnées philosophiques de David Abram et l'argument pour une fondation philosophique de l'animisme. *Les Cahiers philosophiques de Strasbourg*, 49 [Online]. Available at: http://journals.openedition.org/cps/478.

Lalande, A. and Le Marec, J. (2022). Pour une écologie politique des savoirs. Comment hériter du domaine STS ? *Les Cahiers de Framespa*, 40 [Online]. Available at: http://journals.openedition.org/framespa/13069.

Le Marec, J. (2020). Forces des savoirs, précarité des conditions. In *Savoirs de la précarité/Knowledge from Precarity*, Le Marec, J. and du Plessis, H. (eds). Éditions des Archives contemporaines, Paris, 9–34.

Le Marec, J. and Faury, M. (2023). L'enquête, terre d'accueil. Expériences, agencements et conversations. *Communication et langages*, 217(3), 33–52.

Lombardi, D. (2021). Se retrouver, c'est déjà se soigner. Processus thérapeutique, ou de la découverte médiatisée de soi, à travers des pratiques néo-chamaniques. *Cargo*, 11, 65–82.

Losonczy, A.-M. and Mesturini Cappo, S. (2011). Pourquoi l'ayahuasca ? *Archives de sciences sociales des religions*, 153, 207–228.

Luhrmann, T., Nusbaum, H., Thisted, R. (2010). The absorption hypothesis: Learning to hear God in evangelical Christianity. *American Anthropologist*, 112, 66–78.

Mol, A. (2002). *The Body Multiple: Ontology in Medical Practice*. Duke University Press, Durham.

Mol, A., Moser, I., Pols, J. (2010). *Care in Practice. On Tinkering in Clinics, Homes and Farms*. Transcript, Bielefeld.

Morizot, B. (2023). *L'inexploré*. Wildproject, Marseille.

Myers, N. (2018). How to grow livable worlds: Ten not-so-easy steps. In *The World to Come: Art in the Age of the Anthropocene*, Oliver-Smith, K. (ed.). University of Florida, Gainesville.

Myers, N., Kazic, D., Iserte, M. (2020). À propos du Ravissement de Darwin. Entretien Natasha Myers – Dusan Kazic. *La Pensée écologique*, 6(2), 70–76.

Nichols, D.E., Johnson, M.W., Nichols, C.D. (2017). Psychedelics as medicines: An emerging new paradigm. *Clinical Pharmacology et Therapeutics*, 101(2), 209–219.

Sanabria, E. (2021). Vegetative value: Promissory horizons of therapeutic innovation in the global circulation of ayahuasca. *BioSocieties*, 16(3) 387–410.

Servais, V. and Halloy, A. (2013). Divinités incarnées et dauphins télépathes : ethnographie de deux dispositifs d'enchantement. In *Rencontres sensorielles : approches sociologiques et anthropologiques des sens*, Colon, P.-L. (ed.). Petra, Paris, 257–304.

Servais, V. and Servais, C. (2009). Le malentendu comme structure de la communication. *Questions de communication*, 15, 21–49.

Talin, P. and Sanabria, E. (2017). Ayahuasca's entwined efficacy: An ethnographic study of ritual healing from "addiction". *The International Journal on Drug Policy*, 44, 23–30.

9

Tasting Nature: Transforming the Meaning of Wine-tasting Videos on YouTube

This chapter examines the issue of wine-tasting in YouTube videos in the context of the emergence of sustainable practices in winemaking. These practices notably involve the production of natural wines, which can sometimes undermine enological standards, producing surprising-tasting wines, which can less easily be objectively assessed by a tasting expert. Thus, we see the origin of these standards of modern wine-tasting before giving an example in the person of Konstantin Baum – a wine professional whose YouTube channel is among the most popular of all channels devoted to wine-tasting, in terms of audience numbers. By way of contrast, the study also looks at a much smaller YouTube channel, with a smaller audience, the benefit of which is that it presents admittedly amateur tastings. The methodology of comparative study undertaken here is exploratory and is not intended to be exhaustive. Nor is it intended to reflect the vast body of media devices, which are YouTube videos and channels. Rather, the aim is to reveal the characteristics of amateur wine-tasting and show how it can open the door to other receptions of wines drawn from sustainable practices and, in particular, natural wines in the context of ecology of taste. The challenge, which is raised, in my view, by an ecology of taste, firstly, relates to the status of things – wine, in this case – generally considered to be passive and given to powerful symbolization on the part of the people who consume them. That symbolization tends to be organized by a person's own internal logics rather than by the sensory experience itself. A semiotic analysis detached from the mere textual or symbolic productions, as is facilitated by the semiotics of cooperative transactions, as developed by Manuel Zacklad,

Chapter written by Pierre BESLAY.

can reflect this experience differently. The possibility of altering our symbolic representations from taste to a full sensory experience seems essential from an ecological perspective and opens up new theoretical opportunities for information and communication sciences.

9.1. Objectivating the taste of wine

To my mind, the objectivation of the taste of wine involves two parallel ambitions: to name and to rate the taste of the wine. Here, "naming" means defining the semantic field that can be used to describe the taste, against the backdrop of the ambition of breaking free of the ambiguities and artistic fuzz which has hitherto reigned, opting instead for a set of clear, universally understandable terms grounded in a scientific approach. "Rating" refers to assigning scores to the wines – a very widespread practice in the professional wine world, where there is a very strong universal standard in place: the 100-point scale. These two processes help to establish a semantic and numerical objective assessment of the taste, which is therefore no longer something inescapably subjective, which, like colors, we cannot meaningfully discuss.

9.1.1. *Naming the taste of a wine*

The discourse surrounding the taste of wine as we currently know it, and particularly as practiced by professionals in the field of wine, is relatively recent. From ancient times up until the 17th century, as science historian Steven Shapin points out, completely different terminology was used to describe wine: "There are four tastes of wine: sweet, acute, austere, and milde" (Whitaker 1638, p. 19, cited in Shapin 2012, p. 53). The semantic poverty surrounding taste could be attributable to Aristotle, who felt that the sense of taste was inferior to other more indirect senses such as sight because, in taste, "the substance which is being sensed, or an effluvium emitted by it [...] must be in contact with the sensing organ" (Shapin 2012, p. 54). In Shapin's view, it was not until the 19th century and the investment of interest in the chemistry of wine – and other food products made by fermentation – that scientists began identifying the factors in the tastes of wines, but even then, only at a generic level. In fact, it was in the latter half of the 20th century, in the United States, that we began to see a "standardized framework of vocabulary" (Shapin 2012, p. 81) in terms of the taste of wine. A pioneer in this area was the Californian enologist Maynard Amerine (Shapin 2012,

2016), who, in 1983, published *Wines: Their Sensory Evaluation*, along with Edward Roessler. In that book, they put forward descriptors that largely referred to fruits and vegetables (Shapin 2012). Around the same time, Ann C. Noble, their colleague at the University of California, Davis, developed the "Wine Aroma Wheel" – a tool that is widely known to tasting experts and is still used as a point of reference today. At the same time, in France, the enologist Émile Peynaud, who is sometimes referred to as the "father of modern enology", published *Le Goût du vin* (1980), stating in the foreword that "it is now up to enologists to take control of wine-tasting" (Peynaud and Blouin 2013, p. XI). Peynaud and his colleagues enjoyed such success that today, enology is so closely associated with tasting that, for example, to learn how to taste wine, we take "enology classes". However, to understand his original aim, we need to remember the original meaning of the word – which still holds true for wine professionals: an enologist, first and foremost, is the person who provides advice and guidance to the winemaker or overseer during the vinification process. The fact that a technical specialist in wine production is now becoming necessary in the area of tasting is indicative of the professionalization of the role. Finally, we should mention one more important actor, whom Shapin studies: the American wine critic and taster Robert Parker. Parker further imposed this referential, standardized language by adding a new level of justification: that of the democratic ideal of transparency and consumer protection, inspired by Ralph Nader (Origgi 2007, p. 228; Shapin 2012, p. 85).

The purpose of that language therefore needs to be understood in the context of scientific progress. While it is not, strictly speaking, scientific, it nonetheless tastes like science, as Steven Shapin states in the title of his article on the role of the University of California, Davis, in the objectivization of assessing taste: "A taste of science: making the subjective objective in California" (Shapin 2016). Philosopher Ophelia Deroy stresses the fact that the language of tasting is not subject to science: "In professional tasting sessions and competitions, chemical analysis plays only a secondary role, being used to confirm the judgement of the experts" (Deroy 2007, p. 129). Thus, we must ask, what does this modern language of wine-tasting involve? It differentiates between three types of aromas: the primary, secondary and tertiary aromas. Where do these come from? The primary aroma is "an aroma of fruit, 'fruitiness', characteristic of the cépage"; the secondary aroma "develops during the fermentation process" and finally, the tertiary aroma "develops during aging" of the wine (Peynaud and Blouin 2013, p. 34). To describe primary aromas, for example, it is common to find terms referring to red fruits such as strawberries, raspberries and cherries, and also to white fruits or other vegetables, and sometimes spices. To describe secondary aromas, we might speak of milk, bread, butter or brioche. Finally, for tertiary aromas, we could refer to the taste produced by maturing in wood casks – generally with toasted, vanilla or grilled aromas – and later on, for more aged wines, leather, tobacco and prunes. These various examples offer a good illustration of the referential nature of which Steven Shapin speaks: "We can call this type of vocabulary referential, because the obvious intention is to reliably describe the organoleptic characteristics of the wine by referring to the tastes and smells that really are in the wine" (Shapin 2012, p. 51). In Shapin's view, this stands in opposition to a practice designed to "evoke the sensations of drinking a wine through reference to other esthetic experiences" (ibid., p. 51).

9.1.2. *Rating the taste of wine*

The language of wine-tasting as we know it today has gradually become a recognized standard on the basis of the seminal works of Maynard Amerine in California, Émile Peynaud in France, and then the considerable influence of Robert Parker. Parker helped to objectify the taste of wine, bringing it out of the realm of subjectivity, instituted by linguistic norms that tended to be more poetic and evocative than referential and descriptive. However, this objectivization would only be part-way achieved if we were not to speak about what came along with this shift in the language: the scoring of the taste of wines. While the language around tasting lent a scholarly tone to the objectivity of wine, it is all the more evident when we actually speak of numerical scores. Shapin estimates that this scholarly dimension gives meaning to the scoring of wines:

> [D]e-contextualizing the taste is exactly what we need to do. […] And if we stick with the analogy with school, it is one of the virtues of numerical scoring of wines that it does not require reference to the uncontrollable and variable vicissitudes of the occasions on which one might actually drink a wine (Shapin 2012, p. 78).

Shapin points out that 10-point scoring scales became available in the late 19th century – for example, that of enologist Grazzi-Soncini – but "it was neither elaborated with nor linked to any sort of descriptive language" (Shapin 2012, p. 75). Peynaud and Blouin (2013, p. 136) mention "scoring systems from 0 to 5, from 0 to 10 or from 0 to 20". However, today, it is

Robert Parker's famous 100-point scoring system that has become the standard and is used by the majority of professionals, although not without a number of controversies that continually crop up in the world of wine.

One of the justifications for wine scoring relates to the "globalization of the wine market" (Shapin 2012, p. 82), meaning that wines are increasingly being drunk in countries other than the one where they were produced, and by people who sometimes lack the cultural references to understand those wines, or even to understand wine in general. From this point of view, the standardized language and scoring of wine contribute to what the sociologist Michel Callon terms the "market passiva(c)tion of goods" (Callon 2017, p. 56): they help to make these wines marketable goods, which are better able to circulate on the globalized wine market.

Figure 9.1. *Presentation of the app Vivino and bottles of wine about to be tasted by Konstantin Baum. For a color version of this figure, see www.iste.co.uk/souillard/emerging.zip*

9.2. Filming wine-tasting: a video translation of an expert practice

While, at first glance, wine-tasting seems like an experience that would be difficult to share through video, it nevertheless has a strong presence on YouTube. There are numerous channels that have been launched by Masters of Wine[1]; indubitably the largest is that run by German Konstantin Baum, which has over 120,000 subscribers. The largest of all YouTube channels devoted to wine-tasting is incontestably that of the South Korean-dubbed Wineking (YouTube 2017), with over 462,000 subscribers. An honorable mention must also go to the "World of Wine" series of videos (YouTube 2021b) on the channel Bon Appétit, which has over 6.26 million subscribers, dedicated to wine and led by the American sommelier André Mack, some of whose videos have racked up millions of views. All of this is merely the visible tip of a vast iceberg of channels maintained by wine professionals, students and amateurs. In the context of this research, I have focused particularly on Konstantin Baum's channel. Here, I wish to linger over one video that seems particularly emblematic of his channel, in which he tastes wines selling for under $20, which have been scored highly by users of the app Vivino. The video was posted on July 16, 2021 and, although it is not the most-viewed video on his channel, it is still one of the most successful (YouTube 2021b). Vivino is the most widely used of all apps relating to wine (it has over 63 million users). It is used to score, assess and comment on wines based on photos of the labels of the wine bottles. Thus, the promise of the video is that it compares the opinion of a renowned expert to the judgment of an extensive community of amateurs – a lever that creates a sense of intrigue for viewers and thus partly accounts for the success of the video.

Konstantin Baum has demonstrated his expertise over the course of his videos, many of which have mentioned his title as a Master of Wine: for example, "Master of wine tastes the cheapest wines in the world" (2021) or "Tasting like a Master of wine – How to pass the MW tasting" (2021). Others simply use the word "Master" – an incomplete but sufficient abbreviation of the prestigious title. Finally, he has proven his abilities by blind-tasting, comparing glasses of wine or recommending interpretations – all demonstrations of his authority in the subject. The setup in the chosen video is fairly simple: facing the camera, filmed sitting at a table, Konstantin Baum is in the same environment as in the majority of his videos: a stone-built room containing bottles of wine in wooden boxes, crates of

1 The title, Master of Wine, is one of the most prestigious in the world, attesting to tasting capabilities and general expertise about wine.

wine, with tools and accessories relating to wine against the wall. That space and those objects create "numerous modes of presence, scene-setting, and evocation of wine" (Jeanneret and Souchier 1999, p. 78). Over the course of the video, the camera image sometimes briefly alternates with that taken from a different angle, but it does not change the direction of Konstantin Baum's gaze, which remains fixed on the main camera. Above all, this appears to be a strategy to make the video less monotonous. The introductory shot shows Konstantin Baum with all the bottles of wine he is going to taste in front of him, before a short generic speech. For the purposes of introducing the app, the video includes a few smartphone screenshots to show how Vivino works. Finally, the tasting begins. For each bottle, it takes place in the same way: Konstantin Baum begins by presenting the wine, showing the bottle in close-up to another camera; then the image snaps back to the main camera, showing him uncorking the bottle beside him, alongside a glass. From one wine to the next, we see him pouring the wine into the glass, observing its visual appearance, raising it to his nose, and then tasting it[2], but these steps are never shown in full for each wine: the majority are cut into a montage, showing only one or two steps for each bottle, which leads to his commenting on a wine that we have not necessarily seen him drink. Similarly, although it is never shown on screen, we can assume that as an experienced taster, rather than swallowing the wine, he spits it back out into a special recipient. Finally, after, or sometimes between, these various steps, he analyzes the wine in the sensory order of the tasting: first, its visual appearance, then its smell and finally its taste. The visual aspect – the "robe" of the wine – is not addressed for each of these wines. Finally, each tasting concludes with a score on a 100-point scale being awarded to the wine. To summarize, Baum's comments on the first three wines are as follows:

– First wine: Tement Sauvignon Blanc Kalk & Kreide DAC, from the Südsteiermark in Austria:

- robe: bright;

- nose: aromatic; lemon zest, grassy, fresh, green notes;

- mouthfeel: very fresh, very vibrant, beautiful acidity, really good; very precise, it does not really have the complexity;

- score: 90/100.

2 This adheres to the conventional protocol of wine-tasting, with the order: robe, nose and mouth – i.e. visual analysis, followed by olfactory analysis, and finally tasting.

– Second wine: Schloss Johannisberg Gelblack Riesling from the AOC Rheingau in Germany:

- nose: Riesling in its purest form, lemon zest, green apple and also a little bit of petrol;

- mouthfeel: wine is not concentrated, but it has a lot of power and freshness; the acidity really carries the sugar, fruity, lively, vibrant;

- score: 91/100.

– Third wine: Riesling Rotschiefer Kabinett from the Van Volxem estate in the Moselle AOC:

- nose: less intense, more refined, flavors of white peaches, some apple flavors as well, very delicate;

- mouthfeel: fruity, less sweet despite the residual sugars;

- score: 88/100.

It is no coincidence that the comments, thus summarized, take the form of classic tasting notes, as we might find in wine magazines[3]. Konstantin Baum, as a professional taster, precisely applies the method and vocabulary of tasting. Indeed, this is precisely what is most promising about the video presented above: professional tasting wines selected by a broad community of amateurs. Admittedly, the professional tasting is staged here and would look somewhat different outside of the context of a YouTube video. Watching this video, we see an idealized, staged version of documentation of taste, which is at the heart of the world of wine, and without which it would run much less smoothly. What do we find in the video, though, above and beyond the tasting notes? Relatively little: extremely generic elements of context and presentation, a few rare personal judgments and some even rarer phrases which, in contrast with Konstantin Baum's professional stance, belong to the "regime of proximity" studied by Samuel Coavoux and Noémie Roques (2020, p. 185) in relation to video game intermediaries on Twitch and YouTube. For example, we can point to the catchphrase with which he signs off from each of his videos: "Stay Thirsty!". Konstantin Baum's stance is highly professional and relatively aloof, and that is exactly what the Web users watching his videos are looking for. The strength of this video, and indeed of many of the videos on his channel, is precisely in the combination

3 See: https://www.cavepurjus.com/fr/content/fiche-de-degustation.html.

of professionalism with the tastes of amateurs (in the strict sense of people who are not professionals in wine). However, as we will see, wine-tasting on YouTube can be presented in a very different form.

9.3. Filming the taste of wine in a different way: is it an intransmissible experience?

9.3.1. *Presentation of the YouTube channel "ON Y CONNAIT RIEN"*

Our study also looked at a much smaller YouTube channel devoted to wine-tasting – a channel with 871 subscribers, named "ON Y CONNAIT RIEN" (We Know Nothing About It), established in 2021, whose most viewed videos tend to have between 1,000 and 2,000 views (with two videos having over 2,000 views). This channel, presented in French, is maintained by wine-lovers who purposely label themselves as amateurs through their channel's title. In actual fact, the filmed tasting sessions do sometimes involve wine professionals – winegrowers, wine merchants and sommeliers – but the tone is generally that of an amateur setting, where the main selling points are proximity and horizontality, with the presenters being on the same level as the audience. The majority of the videos take the following format: one of the participants in the tasting session, alone, presents the chosen bottle of wine to the viewers; then, they cover it with a blind tasting sleeve and offer a blind tasting to the others, who have to try to guess the cépage[4] or cépages. The original presenter then holds the camera – which is therefore always moving – and films the tasting session while interacting with the other participants. Thus, the viewers can see the participants trying to blindly detect the cépage of the wine while themselves knowing the correct answer. The participants try to guess it, but they do so in a spirit of lightheartedness, as is suggested by the very name of the channel for which the videos are made. Here, the "regime of proximity" is in full effect. The videos are filmed in various settings, generally at the home of one of the participants in the tasting who is hosting the others, in a sitting room, a dining room or sometimes even in the kitchen while a meal is being prepared. Thus, the participants are led to comment on the tasting, and the words they use often fall outside of the "standardized referential vocabulary" studied by Steven

4 The *cépage* is the variety of grape used for the wine. For practical reasons, wines for tasting are often single-cépage (i.e. made using a single variety of grape). When the wine being drunk is a combination of more than one cépage, there is generally one that predominates and can therefore be detected.

Shapin. This does not mean, however, that the tasting does not conform to a framework and rules which owe a great deal to modern tasting standards: olfactory analysis prior to tasting, recognition of the cépage, hiding of the bottle and its label, modern tasting glasses, etc. Finally, each video closes with the revelation of the bottle and comparison of reality with the attempts to guess the cépage.

9.3.2. *Study of an amateur tasting on YouTube*

Figure 9.2. *The presentation of the tasting by Julien Frappa and visual assessment of the wine. For a color version of this figure, see www.iste.co.uk/souillard/emerging.zip*

The choice of one video out of the 151 that were online at the time of writing is more difficult to justify than that of Konstantin Baum's channel. The video chosen (YouTube 2022a) had slightly fewer than the average number of views for a video on the channel. It shows the tasting of a wine from the estate bought by famed French singer and poet Léo Ferré in Tuscany. The video is shot in the kitchen and filmed by Alex – one of the channel's main figures. The bottle is chosen by a winegrower, Julien Frappa, who features in many of the channel's videos. These elements help to put the spotlight on certain characteristics which are also found in all of the channel's videos: Julien Frappa, when he asks the participants to guess the artist behind the wine, and by the way in which he leads the tasting, drives the participants to express themselves in registers which are more surprising and further removed from the conventional descriptors. While Frappa is a wine professional himself, in my opinion, the stance he adopts for this tasting seems to be one of an amateur: he never calls on objective knowledge about tasting – or at least, only occasionally, with no impact on the tasting session itself. Rather, he attempts to bring out a broader range of subjective expressions. Thus, he is playing the channel's game, playing the role of an amateur as suggested by the channel's title.

By contrast with Konstantin Baum's videos, it would be impossible to summarize what is said, in this case, in the form of tasting notes. From the written transcription of the video, we have picked out a few relevant passages which are presented here. First, we have Julien Frappa's remarks framing the tasting session: "focus first on the cépage, and then we'll forget all about everything we've just said, and start on what we can actually detect with our senses" (2022, 2'12"). "If it was an artist, was it a mainstream, civilized artist, or was it a protest artist?" (2022, 3'43"). "Sign in – there is a message in this wine!" (2022, 5'44"). In response to this guidance, the participants speak about the wine in original terms: "Protest artist, but fairly straight as well" (2022, 3'47"); "This is a pop wine!" (2022, 4'39"); "Personally, I really like it. You feel good, it fills me up, and at the same time, I can feel it trying to tell me something" (2022, 5'20"); "It's very left, it's very engaged" (2022, 7'26") and "It's an authoritarian wine, I reckon" (2022, 7'35"). Here, we can clearly see that we are not dealing with the conventional language of wine-tasting but rather with ways of interpreting wine which seek to "evoke the sensations of drinking a wine through reference to other esthetic experiences" (Shapin 2012, p. 51). The governing framework, notably in terms of language, surrounding the tasting here is conducive to a very sensory expression, which is not constant and linear. The terms employed construct an imagine of an active wine, which does or expresses something to the drinker. The semiotician Audrey Moutat, who is a specialist in the language of wine tasting, points out that, even in the context of a classic tasting, the referential descriptors that Steven Shapin so values are not the only acceptable terms, but we can indeed find "so-called 'experiential' descriptors which refer to the structure of the wine" (Moutat 2019a, p. 3). However, these experiential descriptors generally use specific images: "relating to the container […], to architecture […], or relating to human beings" (ibid., p. 1). In other words, the political and artistic images mentioned above fall outside of the conventional language of tasting, to the extent that they could be criticized for what Audrey Moutat identifies as a "rhetoric of transgression" (Moutat 2019b, p. 70).

However, if we have looked only at what the participants say, there is a great deal of content in this video remaining to be analyzed: we see them, in turn, pour themselves some wine, observe its appearance, swirl the glass to aerate the contents, smell it, raise a toast, drink and then throughout the video, periodically come back to one or other of these steps. They are also seen laughing, singing and more rarely, eating. To give readers an idea of how the tasting session proceeds, there follows an extract from the transcript, which also includes the gestures involved in the tasting and the visual cues – all of which seem to be a significant part of the tasting, quite apart from the language – which are connected with it:

> Julie: Already, I feel like I want to toast. Look at these four glasses coming together to pay tribute to the artist.
> *Julie raises her glass to her nose.*
> Matthieu: Hang on, though … I've never drunk a wine with a robe like this. It's really bright!
> *Matthieu observes the wine's color and raises his glass to the light; Julie does likewise. Alex, filming, focuses on his own glass.*
> Julien: Okay – I should point out, though, this doesn't have an appellation. It's a wine, that's all.
> Matthieu: A wine from France.
> *Julie raises her glass to her nose. Julien swirls the wine in his glass, and then drinks some.*
> Matthieu: It makes me think of rum.
> Julie: Well, yeah, absolutely.
> *Julie raises her glass to her nose.*
> Matthieu: It reminds me of rum. When I smell it, the maceration isn't immediately apparent.
> *Matthieu raises his glass to his nose.*
> Julie: It's true, it smells like a digestif.
> *Julie swallows some wine; Julien swirls his glass.*
> Matthieu: Oh, yes, that's very tannic, that takes you by surprise!
> *Matthieu points to his mouth to illustrate the effect of the wine.*

Julien: It is tannic, but that's what you get with orange wines.
It's called an orange wine, hence the color.
Julien displays his glass. Matthieu turns to the camera with an agitated expression, as if to show the effect that the wine has had on him.
Alex: Mathieu, do you like the sensation or not?
Matthieu: Yes, I like it.
Alex looks into the camera.
Alex: Ah, you do like it. He's starting to like macerated wines.
Matthieu: No, but it's not … […]
(YouTube 2022b, 1'20"–2'07")

The majority of these gestures are in reference to the sensory trilogy of wine-tasting which we saw above, with Konstantin Baum: the robe (sight), the nose (smell) and the mouth (taste). Here, though, the amateurs, although their discussion is based on the distinction between the three aspects, do not stick rigidly to a protocol: they are constantly shifting back and forth from one to another.

At this stage, while the transcription is more exhaustive, it does not reflect everything that happens during the tasting session, and what is likely to be of interest to the study: specifically, it does not reflect the effect that the wine has on the guests. Of course, we see various examples of it – for example, through the verbal and physical expressions of pleasure which appear in the passage above – but that is all. However, something does indeed happen, and the wine does something to the group, notably through the alcohol, which influences the guests and their relationship to the experience of taste: this becomes more and more expressive as the video goes on. For example, Figure 9.3 shows what we can glean from the video in addition to the mere tasting.

Figure 9.3. *Timeline of non-verbal expressions over the course of the tasting session*

Here, it is not a matter of painting the guests as "acted upon" by the alcohol, as is suggested by the dichotomy between a wine taster, who is acting, and an intoxicated drinker, who is acted upon[5]. The various expressions found do not originate from the alcohol, but from a multitude of conditions, beginning with the conviviality of spending time with friends, the mindset created by the playful challenge, the theme around art and singing, and it is only in this context that the alcohol is able to help bring about the effect that we see. Here, alcohol is only one easily identifiable example among the many active aspects of wine. It should also be noted that the video contains a number of more educational passages when Julien Frappa explains how the wine has been made.

9.3.3. *The limitations of expert tasting*

The contrast between wine-tasting by experts and amateur tastings has been explored in great depth by sociologists Antoine Hennion and Geneviève Teil. The attempts at objectivity in terms of tasting do not appear to hold water: "the model of expertise as an embedded and objectivized revelation of the qualities of the object is beginning to fail" (Hennion and Teil 2004, p. 5). It seems untenable to reduce the taste to the mere chemical components of the wine: "There remains the same gulf, between the precision of the analysis of the wine's components, on the one hand, and the difficulty in stably translating those measured properties into a perceived 'taste', on the other" (ibid., p. 5). Experts, who are slaves to their analytical model, have no more chance of liking wines "by tasting ten different ones in turn, only to score them and then spit them out, [as they do of liking] music by playing a string of short snippets and cutting each one off almost immediately" (ibid., p. 8). Conversely, amateurs give wine "time and space to do its work" (ibid., p. 9). While in Hennion and Teil's view, blind tasting is emblematic of "objectivist reductionism" (ibid., p. 9), the amateurs behind ON Y CONNAIT RIEN are operating with the attitude which the authors describe: they give the wine what it needs to "do its work" because "just like in music, it is important that there be a shift, the passage of time, exposition" (ibid., p. 8). In contrast, Konstantin Baum seems to us, more, to be a prisoner to the limits of expert

5 Here, we can point to the introduction of the collection "Addictions: plaisir, passion, possession" [Addictions: pleasure, passion and possession], published by Le Manuscrit, on the publisher's website: "It is that tipping point, the moment of changing from acting of one's own volition to being acted upon, from the quest for extraordinary sensations and experiences by a free subject to resignation to dependency by a reified sick person, which we wish to focus on here".

tasting outlined by Hennion and Teil: all indications are that he judges the wines by drinking only a small amount, which he probably does not swallow (although other videos on his channel speak more about the pleasure of tasting).

9.4. How to taste "natural" wines

Up until this point, we have spoken of wine in an indeterminate way, as do Antoine Hennion and Geneviève Teil, and also Steven Shapin and Émile Peynaud, who at most distinguish between good and bad wines, but "if there are bad wines, it is because there are bad drinkers" (Peynaud and Blouin 2013, p. IX). However, it does appear to be true that certain wines, in particular, create remarkable reactions in professional tasters: "natural" or "nature" wines, indeed, "have this propensity to destabilize all the senses and stimulate them by contrast with what is identified (wines stabilized with sulfur, and other exogenous products)" (Pineau 2019, p. 197). The expert taster, Konstantin Baum, has made multiple videos devoted to natural wines, and the way in which he talks about them in a tasting video offers an immediate indication of how a professional taster views them:

> Overall, I do not really care how natural a wine is. [...] I do also care about the fact whether a wine is flawed, and for a long period, natural wines often contained flaws. (Baum 2022, 2'25")

Identification of flaws in the wine is a recurrent theme in the field of professional tasting. Thus, Peynaud and Blouin devote many pages to these flaws in *Le goût du vin*. In particular, natural wines are often accused of containing flaws due to the absence of certain techniques for controlling and preventing bacterial activity, such as filtration and sulfitation. Here, the question of such flaws is not of interest directly in itself, but rather, as an indicator of the attitude of the professional taster – an attitude that often consists of listing good and bad points. However, in the tasting session that he devotes to natural wines, Konstantin Baum does not actually identify any flaws, and appreciates all the wines that he tastes, to the point of declaring himself "pleasantly surprised that there were very little flaws in those wines" (16'08"). On the contrary, the amateurs behind ON Y CONNAIT RIEN are not so afraid of natural wines and their flaws; in actual fact, natural wines represent a high proportion, if not the vast majority, of wines that they drink for their tasting sessions – an obvious example is the Blasfemo from winegrower Mathieu Ferré, which they drink in the video studied above. It seems that it is no accident that we see natural wines so well received by amateurs, in contrast with the suspicion with which they are viewed by expert tasters.

9.5. Semiotic analysis of the tasting of natural wines

Antoine Hennion, addressing semioticians, asks to what extent semiology[6] is capable of dealing with the question of taste, stating that it is a matter of the status of things:

> The stumbling block – and this is the problem around which I will center this discussion – is the status given to things. Do we need to integrate other objects into semiology, bringing still more under the umbrella of "everything is language"? Or conversely, do we need to bring semiology out of the universe of signs whilst maintaining the tools that it has produced, to endow the objects themselves with a capacity to propose, address, call, in short to give them a voice in the chapter? (Hennion 2019, p. 56).

This status of passive objects in the experience of the taste of wine is supported by practices in expert tasting, where wine is viewed as a passive object whose various features are merely waiting for the taster to distinguish them. Indeed, it has been concretely implemented in wines, influenced by these criteria, with the wines being formulated increasingly to satisfy these expectations. In that respect, natural wines are set apart, and in a manner of speaking, reject their assigned place. This has two consequences: in certain professional tasters such as Konstantin Baum, this leads to a form of mistrust as to these wines' ability to respect the specifications for tasting (i.e. the absence of flaws). Conversely, in certain amateurs such as those who run the channel ON Y CONNAIT RIEN, it leads to an experience which is interesting, because it is surprising: these tasters receive a return on their investment of attention in the wine, because it is constantly surprising them and revealing new facts. Hennion expresses this well:

6 In this text, we will use the terms "semiotics" and "semiology" interchangeably (which could be used to distinguish between Peircean and Saussurean schools of thought). The latter is the term used by Hennion.

> We do not inherently like things; something makes us like them. It is the things themselves which reach out to us, but we have to be receptive to that reaching out. Put differently, the lover of something and the thing which he loves make one another: the question is less about what we do than it is about what is being done (and what it is being done to); throughout this long and winding process, made up of revelations and lassitude, unexpected enthusiasm and lasting passions, it is not a matter of gaining mastery over something, but of responding to its call; in turn, provoking these things which are in the process of happening, and allowing them to happen to us – but not only this: in the other sense, we need to revive them and to revive ourselves, and in order for that to happen, we need to have created the space for common sharing and the material organization in which they are arranged (Hennion 2019, pp. 61–62).

Semiotic reflection on taste has become a classic debate in information and communication sciences (ICS), notably thanks to the work of Jean-Jacques Boutaud, who helped to open up the world of semiotics (Boutaud and Veron 2007) to sensory experiences. Thus, in ICS, we no longer affirm that "everything is language", as Hennion decries, and we try to leave room for sensory experience. However, in the case of wine, that place "is not restricted to the moment of consumption" (Boutaud 2016, p. 116). In fact, more emphasis is placed on what leads up to that moment, and what follows it – i.e. the four phases identified by Boutaud: "the prefiguration of the experience, how we prepare for it, how we imagine it; the configuration of the experience, which may or may not conform to those expectations […]; the figuration of the subject entering into this framework – whether they fully and willingly partake of the experience, or remain a little removed from the tasting scene […]; and finally, the reconfiguration of the experience, through any type of translation, communication, discursive or visual recounting of it" (Boutaud 2016, p. 116). Boutaud's major strength, in my view, is that he further enlarged the field of semiotics, making more room within it for sensory experiences. Reading his work, we can see that not everything *is* language: the practices themselves are included in the different phases that he identifies. However, from there to saying that objects themselves are imbued with the capacity to communicate, there is a leap which Boutaud, I believe, is not prepared to make, refusing to be restricted solely to "the moment of consumption", or completely sidestepping the moment of consumption to spotlight, in the different phases, the humans' interpretation of that sensory experience. In my view, the limitations of this approach need to be highlighted from an ecological perspective. Indeed, the anthropologist Léo Mariani, whose research focuses on an Asian fruit, the durian, shows how certain symbolic representations circulating on the subject, particularly in the western world, form an obstacle to a true possibility of discovering the durian (Mariani 2022). These representations are generally characterized by emphasis on the supposedly awful smell that emanates from durians. To describe, in semiotic terms, the authentic discovery of the taste of the durian, Mariani uses the register of indiciality, taken from Peircean semiotics – i.e. a level at which the sign is independent of human representations and conventions, and is based on a connection that already exists between the sign and what it represents. With his book, Mariani warns of the risk of proliferation of symbolic representations which can actually be a barrier to a true sensory experience and, more generally, multiple experiences at the indicial level, of which our relationship to the world is built. We are then faced with the issue of tearing down certain representations in order to make space for a new ecology of taste. This issue, I believe, is particularly difficult to address from Boutaud's perspective, for whom there is an irreconcilable "semiotic disconnect" (Boutaud 2013, p. 85) where "the esthesic dimension of taste and culinary sensation" is constructed in "absence, lack – in this case, lack of flavor and of taste sensations themselves" (Boutaud 2013, p. 86).

Thus, in order to construct an ecology of taste by linking the representations that we have to the actual experience of taste, we will turn to another framework used in ICS, which was originally less oriented towards taste and sensory experiences, but which I believe can help to illustrate it more clearly: the semiotics of cooperative transactions (SCTs), developed by Zacklad (2013, 2020). It combines Greimasian semiotics – while, as Hennion does, rejecting the idea that "everything is language" – with Dewey and Mead's pragmatism. SCT reflects the different actants involved in the transaction – here, the transaction of taste – and their different roles. For a precise definition of actantial roles and a presentation of the transactional framework, refer to Zacklad (2020). In the present case, I denote the guests and leaders of the YouTube channel as the orderers of the wine-tasting, because they created the framework in which they asked Julien Frappa, the entrepreneur, to suggest a bottle of wine, and also a framework for the tasting. Finally, the guests themselves and the wine would, at once, be the realizers and beneficiaries of the tasting. The wine, here, is the realizer because it actively contributes to the tasting, and also the beneficiary, because its qualities can only be realized due to the attention paid to it, as Hennion puts it so well: "It is less a matter of developing the amateur's tastes and the qualities of the object, as two interconnected realities, than maintaining the very relationship that gives life to both of them" (Hennion 2019, p. 62). It is the active and living nature of the wine which is stressed, as it emerges from the actual discourse of the guests, who in turn claim that the wine is "protestative" and "authoritarian". There is no place for such imagination in Konstantin Baum's world, because the wine, which is viewed as inert, is broken down into its various organoleptic components.

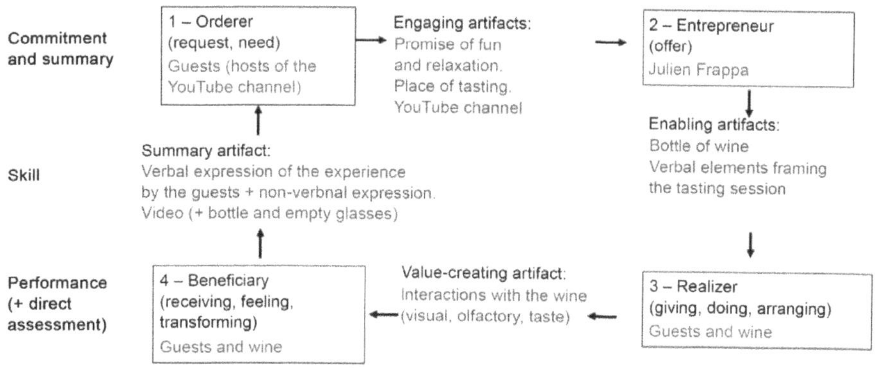

Figure 9.4. *Diagram interpreting wine-tasting using the framework of SCT. For a color version of this figure, see www.iste.co.uk/souillard/emerging.zip*

The main idea underlying SCT is that any transaction is mediated by one or more artifacts, which may be either symbolic or tangible. This allows us to highlight the role of each of the actants here, and what they rely on in order to allow the transactions of tasting to take place. While certain aspects may seem trivial, such as the "empty glasses", indicating that the wine has been drunk, they set this tasting session apart from that of Konstantin Baum, who probably did not finish the glasses of wine (much less the bottles) chosen by the Vivino users, for the simple reason that he did not enjoy them at all. The transactional dimension is all the more relevant if we originally refer to the notion put forward by John Dewey and Arthur Bentley (1949), because they contrast *trans*action with *inter*action, the latter being a context in which action occurs between two parties who are not, in their turn, affected by the action. Yet this is exactly what happens in the context of professional wine-tastings, as we can see in Konstantin Baum's videos: the professional must remain unimpaired to employ his analytical expertise and must stably establish the identity of the wine by breaking it down into its different organoleptic components. There is no place for such stability, objectivity and aloofness from a transactional perspective, where the two poles are interlinked, and co-evolve as the transaction takes place: the wine is gradually revealed and evolves, and the guests passing from one state to another as the alcohol takes effect. With professional tasting, the "semiotic disconnect" of which Jean-Jacques Boutaud spoke is in full effect: we need to look at the symbolic traces to reconstruct the actual experience of the taste, if it is indeed possible to do that, if such a "semiotic suture" (Parret 2006, p. 150; Boutaud 2013, p. 85) can be made. With the amateurs, it is different: the wine, with its identity less firmly established, is drunk and takes its effect. The representations that it triggers are articulated with the visual, olfactory and taste indicators that are constantly revisited by the guests who, while they struggle to establish a stable and objective identity for the wine, discover its nuances and subtleties and express them not only in unconventional language but also through gestures, singing and laughter.

9.6. Conclusion

When tasting the natural wines, Konstantin Baum was "pleasantly surprised". This appears to indicate that these wines' "propensity to destabilize" fortunately worked, despite the fact that the expert taster did not give them the time and space that wine drinkers would. Could it be that the same symbolic saturation that prevents westerners from enjoying the taste of durian fruits similarly creates a barrier to the taste of natural wines? Léo Mariani does not say that westerners never eat durians, but rather, when they eat the fruit, their pre-existing mental representations of it prevent them from fully enjoying the experience, because they are only able to confirm their ideas concerning the smell of durians. Thus, it is a certain experience of the world which remains inaccessible to them. Konstantin Baum's surprise is undoubtedly a good thing, but it also demonstrates the extent to which certain representations are rooted in the mind of a professional taster and destroy his appreciation of natural wines. However, to speak only of representations would be to ignore the concrete practices in which they are embedded: the strict protocols governing tasting which, while they do effectively establish the objective character of the wine, allows it very little chance to reveal its full complexity, beyond the series of expected organoleptic markers. Rather, it is in the amateur tasters' videos that we actually see these wines take their full effect: not being restricted by professional tasting standards, they can give the wines all the time and space they need to do what they were made to do. Not all wines need time and space, because today, the majority are made to conform to the representations circulating among professional tasters – as though durian growers had bred varieties without their expected unpleasant odor. From this point of view, natural wines are an ecological option which not only aims for a more restricted use of the techno-chemical solutions developed in viticulture and vinification but also an attempt to reopen the field of taste, and make it more ecologically friendly. The YouTubers, showing two completely divergent possibilities, create their videos in the same way as vintners create wines: either as an exercise in mastery, producing an expected, linear and standardized result, or as a freer exercise that can produce a more chaotic and sometimes disquieting result, but which is still rich and lively. In the latter case, what we begin to see is what Hennion and Teil (2004, p. 7) term "the work of wine": "What it does, what it produces, the pleasure and the sensations that it can trigger, the way in which it takes, seduces, and moves, based on the object wine". Thus, we see the possibility that communicational devices could be closely interlinked with the sensory experience in order to help it manifest fully, with all its richness, despite creating a little disorder in the symbolic representations of the wine.

9.7. References

Boutaud, J.-J. (2013). L'esthésique et l'esthétique. La figuration de la saveur comme artification du culinaire. *Sociétés et Représentations*, 34, 85–98.

Boutaud, J.-J. (2016). Le vin et l'éveil des sens. L'expérience du goût en partage. *Hermès, La Revue*, 74(1), 110–119.

Boutaud, J.-J. and Veron, E. (2007). *Sémiotique ouverte. Itinéraires sémiotiques en communication*. Hermès-Lavoisier, Paris.

Callon, M. (2017). *L'emprise des marchés : comprendre leur fonctionnement pour pouvoir les changer*. La Découverte, Paris.

Cave Pur Jus (2013). Fiche de dégustation Vin – Télécharger gratuitement [Online]. Available at: https://www.cavepurjus.com/fr/content/fiche-de-degustation.html [Accessed 15 December 2023].

Coavoux, S. and Roques, N. (2020). Une profession de l'authenticité : le régime de proximité des intermédiaires du jeu vidéo sur Twitch et YouTube. *Réseaux*, 224(6), 169–196.

Deroy, O. (2007). The power of tastes: Reconciling science and subjectivity. In *Questions of Taste, The Philosophy of Wine*, Smith, B. (ed.). Signal Books, Oxford.

Dewey, J. and Bentley, A.F. (1949). *Knowing and the Known*. Beacon Press, Boston.

Hennion, A. (2019). La force des choses : le goût comme engagement. Politiche del gusto. Mondi comuni, fra sensibilità estetiche e tendenze alimentari. In *Actes du 46e congrès de l'AISS*, Palermo, XIII(25), 55–69.

Hennion, A. and Teil, G. (2004). Le goût du vin : pour une sociologie de l'attention. In *Le goût des belles choses. Ethnologie de la relation esthétique*, Nahoum-Grappe, V. and Vincent, O. (eds). Éditions de la Maison des sciences de l'homme, Paris.

Jeanneret, Y. and Souchier, E. (1999). L'étiquette des vins : analyse d'un objet ordinaire. *Communication et Langages*, 121(1), 72–85.

Le Manuscrit (2012). Addictions : plaisir, passion, possession[Online]. Available at: https://lemanuscrit.fr/collections/addictions-plaisir-passion-possession/ [Accessed 15 December 2023].

Mariani, L. (2022). *Le goût des possibles. Enquêtes sur les ressorts symbolistes d'une crise écologique*. Presses Universitaire de Paris Nanterre, Nanterre.

Moutat, A. (2019a). Le commentaire œnologique ou comment déguster le vin par les mots. *Lexis. Journal in English Lexicology*, 13.

Moutat, A. (2019b). Créativité et rhétorique des discours promotionnels sur les vins naturels. *Recherches en Communication*, 48.

Origgi, G. (2007). Wine espistemology: The role of reputational and ranking systems in the world of wine. In *Questions of Taste, The Philosophy of Wine*, Smith, B. (ed.). Signal Books, Oxford.

Parret, H. (2006). *Sutures sémiotiques*. Lambert-Lucas, Limoges.

Peynaud, E. and Blouin, J. (2013). *Le goût du vin : le grand livre de la dégustation*. Dunod, Paris.

Pineau, C. (2019). *La corne de vache et le microscope : le vin "nature", entre sciences, croyances et radicalités*. La Découverte, Paris.

Shapin, S. (2012). The tastes of wine: Towards a cultural history. *Rivista Di Estetica*, 51, 49–94.

Shapin, S. (2016). A taste of science: Making the subjective objective in the California wine world. *Social Studies of Science*, 46(3), 436–460.

Teil, G. (2001). La production du jugement esthétique sur les vins par la critique vinicole. *Sociologie du Travail*, 43(1), 67–89.

Whitaker, T. (1638). The tree of humane life, or, the Bloud of the Grape. Proving the possibilities of maintaining humane life from infancy to extreame old age without any sicknesse by the use of wine [Online]. Available at: https://openlibrary.org/books/OL7013779M/The_tree_of_humane_life_or_The_bloud_of_the_grape.

Wikipedia (2023). Émile Peynaud [Online]. Available at: https://fr.wikipedia.org/w/index.php?title=%C3%89mile_Peynaudetoldid=200118672 [Accessed 15 December 2023].

YouTube (2017). Wineking [Online]. Available at: https://www.youtube.com/@wineking [Accessed 15 December 2023].

YouTube (2021a). Master of wine tries highly rated wines from vivino under $20. Konstantin Baum – Master of Wine [Online]. Available at: https://www.youtube.com/watch?v=yju8sTp78BY [Accessed 15 December 2023].

YouTube (2021b). World of wine with André Mack, Bon Appétit (playlist) [Online]. Available at: https://www.youtube.com/watch?v=A79U_-qnMmYetlist=PLKtIunYVkv_RpjXAgcKGgR2BFXpgBU7FQ [Accessed 15 December 2023].

YouTube (2022a). Are natural wines good or bad? Master of wine tastes natural winese. Konstantin Baum – Master of Wine [Online]. Available at: https://www.youtube.com/watch?v=6ghGYvUoKAQ [Accessed 15 December 2023].

YouTube (2022b). Dégustation Vin – San Donatino – Petit Manseng – Blasfemo – Mathieu Ferré – Dégustation à l'aveugle, On y connait rien [Online]. Available at: https://www.youtube.com/watch?v=LGcY0eMdl2E [Accessed 15 December 2023].

Zacklad, M. (2013). Le travail de management en tant qu'activité de cadrage et de recadrage du contexte des transactions coopératives. *Activités*, 10(1), 192–220.

Zacklad, M. (2020). Changements de régimes de conversation dans la transition numérique. *Approches Théoriques en Information-Communication (ATIC)*, 1(1), 7–40. doi: 10.3917/atic.001.0007.

PART 5

Plant Communication as a Means of Transforming Organizational Communication

10

Uncertainty in Strategic Environmental Communication

Strategic communication can powerfully influence how audiences think about their relationship to the natural world. Much environmental communication has been written specifically to inspire feelings of stewardship about the natural world in audiences. Much environmental communication has been written specifically with this goal in mind. Romantic authors like John Muir wrote about nature in near-religious ways and influenced a cultural view of nature as a pure and sometimes divine space (James 2014). The nonhuman and especially vegetal world was often at the center of Muir's (1897) nature writings, which made clear his admiration for the forests of North America. It is important to note, however, that Muir did not write about nature in this way simply for his own enjoyment but often for furthering discrete goals, including advocating for the establishment of a national park system in the United States and for the use of nature as a form of recreation (Muir 1901, 1920). Later nature writers have challenged the legacy of his works and others like him, pointing out the negative consequences of this nature concept, such as the marginalization of native peoples in the name of an arbitrarily defined concept of "nature" (Murdock 2021). The strongest of these criticisms could themselves be seen as strategic in nature, such as Robin Wall Kimmerer's *Braiding Sweetgrass* (2013), which explicitly advocates for multiple ways of learning about the natural world, in addition to Western empirical science. Environmental writings such as Braiding Sweetgrass (Kimmerer 2013) discussed plant life distinctly from such Western-centric nature writings, most conspicuously by

Chapter written by Joshua T.L. ANDERSON, Matthew EASTIN and Megan LAMONICA.

discussing the ways in which humans can sustainably interact with plants, as opposed to removing humans from the natural world.

Strategic environmental communication has been a topic among academics and practitioners alike as a means of building public will to prioritize environmental goals and change behaviors such as consumer purchases and other environmentally relevant behaviors like water conservation (Atkinson 2012; Dalrymple et al. 2013; Liang et al. 2018). The field of strategic environmental communication could be seen as a sub-field of the larger field of strategic communication, with many of its academic foundations borrowing from the broader field of strategic communication (see Fuller 1999). In this chapter, we present strategic environmental communication in a way that may better address environmental challenges by drawing inspiration directly from studies of the natural world.

The central premise of this chapter is that the theories that have allowed researchers to study the ways that organisms such as plants can live and thrive in the natural world can be adapted for the purpose of research-guided strategic communication that could support the natural world. In particular, we will be discussing how the ways in which natural scientists study the uncertainty present in natural environments might inform how we think about uncertainty in communication environments for environmental organizations. In order to conceptualize this theoretical application, we will begin by reviewing the communication concepts central to environmental strategic communication, discuss how biologists conceptualize uncertainty in natural environments and construct an application of their theoretical work to the study of communication environments.

In short, a holistic perspective on environmental strategic communication that is inspired by studies of natural ecology is presented. As part of our proposal for environmental strategic communication, we discuss uncertainty as a feature of communication environments overall, which may not be readily apparent from a study of the actors in the environment. Here, environmental uncertainty will be positioned as a potentially important concept for the planning of communication strategies for environmental organizations and will question how we might draw on studies of biological systems in a deeper way to give us greater insight into communication systems. In particular, drawing upon theories of natural selection to consider how we might conceptualize uncertainty in communication environments and how we might develop strategic environmental communication that can work to reduce or cope with environmental uncertainty.

10.1. Strategic communication to serve environmental goals

Strategic communication can take many forms, including advertising, public relations and other related disciplines (Heide et al. 2018; Werder et al. 2018). These forms of communication are united under the classification of strategic communication because they are all goal-oriented (Hallahan et al. 2007; Zerfass et al. 2018). At its core, strategic communication entails any phenomenon where an entity is using communication as a means to achieve one or more of its goals, which can include behavior or attitude change but can also include less obviously persuasive goals such as maintaining a positive image (e.g. Moyer-Gusé et al. 2019).

While an orientation towards goals is a definitional quality of environmental strategic communication and may be the focus of academic research, the actors that are involved in environmental strategic communication are also seen as important subjects of research. In the broader field of strategic communication, stakeholder theory generally encompasses works that examine the relevant actors with whom an organization engages in communication (Friedman and Miles 2002). In an environmental context, stakeholder theory serves as the basis for studies in corporate social responsibility and environmental social governance, wherein organizations maintain their positive image by promoting themselves as responsible actors (Eccles et al. 2020). Stakeholder theory has environmental applications to non-profit entities as well, as these organizations work to identify key stakeholder groups, including their supporters, their constituents and the individuals that they serve (Crane and Ruebottom 2011). Ultimately, studies of environmental strategic communication not only examine environmental goals that a communicative entity attempts to reach (and if they meet them) but also who they need to communicate with to meet them.

A final subject of research discussed in this brief review relates to how an entity engages in communication, specifically the media channels they choose to adopt. In communication research, the term "channel" generally refers to the medium through which communication takes place; much work in strategic communication has affirmed the importance of considering the specific affordances of channels that an entity might choose to further its communication goals (e.g. Botan and Taylor 2006). Among the most important works for studying channel characteristics is Sundar's (2008) Modality-Agency-Interactivity-Navigability (MAIN) model. The MAIN model suggests that media can be evaluated in terms of the different ways that information can be presented (Modality), the ability for individuals using the medium to give their own input (Agency), the degree to which users engage with an interface in a back-and-forth manner (Interactivity) and the ability for users to move between content on a medium (Navigability) (Sundar 2008). The MAIN model, and other research models like it, has allowed scholars of strategic communication to think of channel-level differences in media forms, which can be used to suggest the relative usefulness of different channels for achieving specific communication goals (Rathnayake and Winter 2018; Sundar et al. 2019).

Taken together, the field of environmental strategic communication is well-grounded in the broader field of strategic communication and takes on much of the broader field's interests in communication goals, stakeholders and media. Environmental strategic communication may also benefit, however, from other theoretical perspectives that originate outside of the broader field of strategic communication. Theories of ecology may be particularly valuable in this regard because of their orientation towards holism, with ecosystems being the primary subject of inquiry (Jørgensen and Mejer 1979). When an ecologist studies an ecosystem, they cannot only look at each organism in the ecosystem; they also observe the ecosystem as a whole. We posit that this approach may have merit for studies of strategic environmental communication, and thus, we propose a way of looking at this phenomenon that is inspired by studies of natural selection. Specifically, research into natural selection may be especially valuable in terms of the insights they give into how organisms can cope with uncertain environments, which may parallel the struggles of environmental organizations in a frequently shifting media environment.

10.2. Environmental uncertainty in biological systems

Natural selection, or the processes that explain differential success of organisms in their respective environments (Brunnander 2007), is based on Charles Darwin's (1859) now foundational writings on the subject. As he attempted to explain the subtle differences in geographically close species that he observed during his famous voyage on the HMS Beagle, Darwin theorized a mechanism of descent driven by gradual species differentiation in response to environmental pressures (Darwin 1859). While Darwin's (1859) most famous writings on his observations were about animal species such as finches and tortoises, his theories on evolution have supplied a foundation to studies across the tree of life, including vegetal life. Natural environments will always favor some organisms' physiologies and behaviors (generally called "strategies") over others, and over time this will lead to the most well-fit strategies becoming more prominent (Brunnander 2007). This general principle of Darwin's (1859) work is commonly encapsulated by the phrase "survival of the fittest"; however, its usage in popular media often misses some key points. First, it is incorrect to think that this process is converging towards the development of a generally superior organism that is ideal in all or most environments, for which the phrase "survival of the fittest" is commonly misconstrued to mean in popular media and by political movements such as Social Darwinism and eugenics (Gordon 2004). Instead, the process is a diverging one; as different environments place different demands on organisms, the same strategy may be more or less effective than other strategies depending on the specific environment. Second, the use of the term survival is often taken to mean that the primary operative function of differential success is avoidance of death; however, it is actually successful reproduction. An organism is not necessarily successful if it is able to live to an old age if it does not successfully reproduce, and alternatively, its strategy is not necessarily a failure if an organism dies young, as long as it was able to leave descendants before it expired.

Darwin's work still has a great deal of influence on the fields of evolutionary biology and ecology, but like all scientific works, it has shown its age as more recent research reveals insights into the process of speciation. One update to Darwinian evolution that is often discussed in popular media is the study of epigenetics, which has revealed that environmental conditions can shape how innate traits may be expressed, and patterns of expression may themselves be heritable (Weinhold 2006). While insights related to epigenetics challenge the long-held assumption of Darwinian evolution that heritable traits are innate to organisms, the assumptions of epigenetic evolutionary models still rest on the body of traditional genetic work, as it deals with the expression of innate and heritable genes within organisms subject to evolutionary pressures in their environments (Weinhold 2006). Another notable body of work for the purposes of this chapter is the theory of r/K selection, which digs deeper into the ways that strategies may be more or less favored depending on the resources that organisms invest into their offspring. The theory of r/K selection was popularized by Robert MacArthur and Edward O. Wilson (2001) in their observations of island ecosystems. Many years after Darwin's (1859) own island studies, MacArthur and Wilson (2001) conducted a wider survey of island ecosystems and found that islands may vary in terms of the stability of life on the island. While some islands provided a relatively stable life for the organisms on them, with relatively predictable rainfall, temperatures and tides, some islands were notably more chaotic (MacArthur and Wilson 2001). While intuitively one might assume that most species would want to live in a stable environment, MacArthur and Wilson (2001) noticed that species who invest relatively few resources into their offspring seem to compete better on islands with more chaotic environments.

Based on this insight, MacArthur and Wilson (2001) discussed the potential to classify most species as either "r selecting" (who invest few resources into their offspring) or "K selecting" (who comparatively invest a good deal of resources into their offspring). The investment of resources, in this context, is variously defined and entails more energy, time and attention being devoted to each offspring by a parent. Differential investment is essentially characterized by r-selecting species producing more offspring than K-selecting species over a given period of time, but there are consequences to investing so few resources into individual offspring. The individual offspring of K-selecting species are more likely to live to adulthood than those of r-selecting species (Cassil 2019). An extreme example of the difference between an r-selecting and a K-selecting species is a fruit fly and an African elephant, respectively. The difference in investment can be so extreme that 66 successive generations of fruit flies (which can be produced as often as 10 days) (Fernández-Moreno et al. 2007) can take place over the 22 months that a mother African elephant takes to gestate one calf (Hildebrandt et al. 2007).

Put simply, r-selecting strategies are favored in more chaotic environments, and K-selecting strategies are more viable in more stable environments (MacArthur and Wilson 2001). Key to understanding this dynamic is understanding that chaotic environments are characterized by deaths occurring due to random chance. When widespread death happens by random chance, these events disproportionately affect individuals that place a comparatively higher degree of investment in each offspring. To return to the prior example, if a fruit fly, which can lay hundreds of eggs in a single clutch, were to lose half of its descendants by random chance, that would not be nearly as devastating as if an African elephant were to lose half of her calves by random chance. The same dynamics also govern the evolutionary pressures of plant life; for example, large trees such as baobabs or California redwoods often place substantial resource investment in their offspring and are disproportionately susceptible to random environmental disturbances. At a population level, K-selecting organisms will struggle to survive in environments that expose them to a high chance of random death, but r-selecting populations can often survive in these environments.

As we work to adapt the insights from MacArthur and Wilson's (2001) theorizing to the field of environmental strategic communication, we must acknowledge the challenge in bridging such distant disciplines. This chapter does not seek to treat communication actors as operating like biological species directly, but instead to understand the theory of r/K as a general guide to coping with chaotic environments, and then seek to use this theory to inform how scholars might think of, and how environmental communicators may best operate within, chaotic or uncertain communication environments. The core idea is that more uncertain environments make higher comparative investment in individual prospects a worse strategy than spreading investment across more prospects than it would be in comparatively less uncertain environments. In the following section, we describe how this general principle may be explored in communication systems.

10.3. Environmental uncertainty in communication systems

The term communication "environment" refers to the full scope of communications that are produced or consumed by an entity, including individuals and organizations typically involved in environmental strategic communication. Individuals and groups engaging in strategic environmental communication must consider many aspects of their communication environment to achieve their goals, such as their target audience and the media channels that they can use to reach their target audience. So what does "uncertainty" mean in the context of a communication environment? We use the term "environmental uncertainty" in the context of communication to refer to an inverse measure of how surely an entity can achieve its communication goals within its communication environment. Thus, we characterize a communication environment that gives few guarantees of success as a communication environment with a high degree of uncertainty.

Importantly, because environmental uncertainty describes the sum of the myriad environmental conditions that affect the chances that an entity's communications will work to achieve its goals, it is impossible to truly estimate in an absolute sense and is more useful as a point of comparison. For instance, X (formerly Twitter) accounts with more followers could be said to have lower environmental uncertainty than those with fewer followers. This is because followers allow a more direct line of communication to people that might be interested in an organization's communications than would be attained by an organization posting and hoping that the communications will naturally spread through their network (Soedarsono et al. 2020). In this example, however, an account can only be said to have lower environmental uncertainty than another with fewer followers when all other contextual factors are held constant. While this condition is infeasible in the real world, it is of interest to study changes in overall environmental uncertainty by studying comparisons of components. Another way of thinking of this comparative view is the changes that can occur in a single entity. Continuing with this example, a social media account on a platform like X can be said to reduce their environmental uncertainty if they grow their follower base on a social media platform, assuming that no other contextual factors change. In addition, we must also acknowledge that not all followers would have an equal impact on uncertainty reduction. Followers with relatively large networks may be especially adept at reducing uncertainty, because they may spread messages relatively far when they share them. Conversely, when follower accounts are bolstered by low-value followers such as bot accounts that do not share their message to a wider network, these may do little in the way of reducing environmental uncertainty. Again, it must be emphasized that environmental uncertainty may only be conceptualized as a rough measure, and it is important that it be contextualized within the specific conditions of a communicative entity.

The current perspective posits that, all other factors being equal, more uncertain communication environments make higher comparative investment in individual communication events a less optimal strategy than spreading investment across more events than it would be in comparatively less uncertain communication environments. Even though there are differences between biological and communication systems in application, which we will discuss below, we argue that this insight inspired by research conducted into the natural world can yield valuable guidance into research and practice in strategic environmental communication. The application of this insight to strategic environmental communication is discussed in terms of the three key considerations of strategic communication already discussed, namely: goals of communication activities, identified stakeholders and media considerations.

10.4. Environmental uncertainty for communication strategy

It would be helpful to have some sense of what an "r-selecting" communication entity is, as opposed to a "K-selecting" communication entity. The cleanest way to think about the question of comparatively high resource investment in a communication context is to consider strategic communicators whose overall strategy is premised on high specialization in terms of goals, stakeholders or media. That is, an environmental organization whose key focus is grounded in pursuing a small number of goals is more specialized than an organization whose key focus is set on pursuing a greater number of distinct goals and thus closer to a K-selecting strategy. The same dynamic could be discussed for organizations that have a comparatively low number of stakeholders or those that employ comparatively few forms of media to communicate with stakeholders. A prototypical example of a strategic environmental communication entity engaging in a K strategy may be Clean Creatives, which describes itself as "a project for PR and ad professionals who want a safe climate future" (Clean Creatives 2023). The focus of communications from Clean Creatives is their landing page-style website, which asks creative professionals (including advertising and public relations professionals) to take a pledge to not work with fossil fuel producers. Overall, Clean Creatives is a highly specialized strategic environmental communication organization, evidenced by their very narrow goals of trying to deny strategic communication talent to fossil fuel industries and their relatively small group of primary stakeholders (strategic communication professionals).

While we can conceptualize the strategies of environmental strategic communicators in terms of r or K strategies parallel to strategies in nature, to continue the connection, we also need to discuss the dynamics of selective pressures that may lead these groupings of strategies to be successful in some communication environments over others. It likely comes as no surprise to practitioners of strategic communication that specialization can be a great asset in achieving communication goals. When an organization such as Clean Creatives knows the stakeholders they want to reach, they have the liberty to tailor their messages in a way that those stakeholders are likely to respond well to, and they may be able to have their intended impact. Message tailoring has proven itself to be a major tool in strategic communication (Hirsh et al. 2012). Message tailoring in these communication environments means that strategic communicators have a lower chance of encountering resistance. A heterogeneous audience of key stakeholders could make opposing demands for message content and framing, but effective message tailoring will facilitate a lower risk of generating a negative response from one audience segment when trying to speak to another (Hanus and Fox 2017).

While many of the advantages for specialization come with tailoring messages for particular stakeholders, there are distinct advantages for strategic environmental entities that have the liberty to maintain a small number of media channels for their communication. Media like newsletters may be useful in serving informational roles with dedicated bases of supporters and may be a valuable lynchpin in a communication strategy that centers on spurring these supporters to action. That said, these media may be less useful in attracting new supporters, especially those that expect more interactive and engaging forms of media (Sundar et al. 2019). While there are many success stories of strategic communication entities being rewarded for agility in their movement between media channels, adaptation among different media channels requires a division of resources and expertise that can be costly. This is especially true in a contemporary media environment that is marked by a wide array of social media platforms, each with their own specific features and affordances, in addition to all other forms of media. Overall, we would argue that the ability to maintain a high degree of specialization in terms of audiences and media should be generally seen as an advantage to strategic environmental communicators that is most often afforded to those who inhabit relatively stable communication environments.

If we assume there are disadvantages to operating in a highly uncertain communication environment compared to a less uncertain environment, a logical next question is how strategic environmental communicators can shape their environments to reduce uncertainty. Beginning with stakeholders, a major source of uncertainty in a communication environment may be situations when strategic communications are ill-fit to key stakeholder groups and incur negative responses (Hanus and Fox 2017). One possible avenue for reducing uncertainty here is for strategic environmental communicators to narrow down their key stakeholders to as homogenous a group as possible while still succeeding in their communication goals. Depending on the goals of the communication entity, this may not always be possible. For the case of strategic environmental communicators like the American Conservation Coalition, their mission of "building the conservative climate movement" (American Conservation Coalition 2023) is necessarily premised upon bridging groups of stakeholders that are heterogeneous in meaningful ways, including political orientations. Other strategic environmental communicators, however, may have a heterogeneous group of stakeholders due to poor connection to their central goals. Such entities may unnecessarily run the risk of communicating in a way that appeals to no one group of stakeholders in particular because they are trying to appeal to too many stakeholders. A similar hazard may also occur if a strategic environmental communicator does not know their audience and is unable to engage in meaningful communication tailoring, even though their core groups of stakeholders are relatively homogenous.

In these cases, the form of environmental uncertainty described above derives less from a chaotic environment and more from a lack of knowledge that the communicator has about their environment. This type of knowledge may also be important for reducing uncertainty related to media. A key piece of knowledge that a strategic environmental communicator can discover about their audience is which forms of media can be used to best engage with them. In addition to being unaware of which media channels best fit with each of their key stakeholder groups, environmental strategic communicators may also encounter environmental uncertainty in the form of a knowledge deficiency regarding possible media channels. Even though a strategic environmental communicator does not employ a particular media channel in their work, the media channel may still be a feature of their wider communication environment. Strategic environmental communicators can thus reduce environmental uncertainty related to media channels by staying abreast of new media channels and understanding how these may (or may not) be effective tools to communicate with their key stakeholder groups.

These forms of uncertainty reduction, however, are premised on a strategic environmental communicator having a clearly defined set of goals that inform strategic choices surrounding stakeholders and media channels. If a strategic communicator has an unclear set of goals, or too many disparate goals, this may be an especially powerful contributor to environmental uncertainty. Strategic communication entities in this situation may find themselves trying to communicate to a vast array of stakeholder groups, using a huge number of media channels. While large communication entities may be able to maintain this approach, it is likely to be costly and will necessarily incur a great deal of environmental uncertainty. Massive communication entities that fall into this category include powerful nation states, whose key communicators have often spoken to the challenges of navigating this degree of environmental uncertainty (e.g. García 2012). Overall, we argue that one of the most powerful uncertainty reduction approaches for strategic environmental communicators is to pursue goals in such a way that their core goals are specific and parsimonious.

There are occasions when communication environments will be characterized by high uncertainty, and uncertainty reduction approaches may not be effective in altering these environments in such a way that favors a highly specialized approach. As discussed above, these situations may result from diverse or expansive core goals, but they may also be situational factors arising outside of the core strategy of a communicative entity, and they may be vulnerable to social change. One such example is when a key stakeholder group fractures along a line that is hard to reconcile, transitioning from a relatively homogenous group to a more heterogeneous group. For example, previously non-controversial topics like willingness to engage in environmental social governance may now be treated as signifiers of identity that alienate segments of the customer base of many of these entities (Zahn n.d.). These entities can no longer tailor their communications to emphasize these traits for stakeholders that respond well to them, including stockholders who have a vested interest in responsible corporate management, without risking backlash from stakeholders that respond negatively.

Another source of growing uncertainty in contemporary communication environments is based on the explosion of new media channels. Digital media channels have not only become more popular over time (Pew 2023) but have also become increasingly diverse. The contemporary social media ecosystem has a great number of smaller platforms with highly engaged bases of users, in addition to the largest platforms. These small platforms may have very small user bases that are often highly distinct in terms of political orientation or other characteristics (Rogers 2020). Many novel forms of media may quickly rise or fall in popularity as well. A salient example of the fickleness of popularity is the massive success, and then failure, of various "Web3" media over the early 2020s, such as collectible Non-Fungible Tokens (NFTs) (Rosen 2023). Ultimately, the contemporary communication ecosystem is highly complicated in terms of media forms, especially, and strategic environmental communicators may struggle to develop effective media plans to reach their audiences. Unless there is a reversal in these trends of digital media, strategic environmental communicators are likely to be forced into coping with this source of uncertainty.

A key question to follow this discussion is what, specifically, strategic environmental communicators can do to cope with uncertain communication environments. The core principle of r-selecting-like communication strategies, which may cope better with uncertainty than K-selecting-like strategies, is they favor breadth of approach over a highly specialized approach. As opposed to neatly tailoring communications to a core group of fairly homogenous stakeholders, they may try to appeal to a broader group of stakeholders that may have a diverse range of environmental interests. Strategic environmental communicators, in some cases, may experience success in finding valuable groups of stakeholders if they notice that some of these communications test well. Likewise, a similar approach can be taken to media investments in environments characterized by a relatively high level of uncertainty. If the most effective media tool is unclear, strategic environmental communicators may find success by spreading out their media investments broadly. While many of these media investments may turn out to be failures, there can be great rewards for success. Reflexivity may be key to successfully implementing this form of strategy, however, in order to parse successful media investments from failed media investments. As a strategic environmental communicator pursues a broad communication strategy in a relatively uncertain communication environment, they may take note of their successes and failures, and their communication environment becomes at least somewhat less uncertain. Overall, while strategic environmental communicators that work within relatively low-uncertainty environments may have an easier time developing highly specialized communications, we argue that communicators operating in more chaotic environments can better their chances of success through uncertainty reduction techniques or by employing strategies characterized by a wide distribution of investment. That said, our argument is highly exploratory in nature and is largely rooted in applications from a distant academic field, which warrants some discussion of limitations.

10.5. Limitations and considerations for application

While we believe our crossing of disciplines is a worthwhile enterprise that could provide valuable insights for the field of strategic environmental communication, we appreciate that our application of r/K selection theory requires a more in-depth discussion of the differences in application of its general principles. One key difference in application is that strategic environmental communicators are engaging in conscious strategies, whereas the term "strategy" in biological studies implies something more like innate traits that are coded into the genetic material of organisms. While a communication organization can opt for uncertainty reduction approaches to try to pivot to a more specialized communication strategy, a patch of Bermuda grass cannot become something like a baobab if it puts in enough effort. The theory of r/K selection was positioned to explain phenomena in evolutionary biology that occur over eons with little conscious input from those they act on. Communication phenomena are, by contrast, much shorter-lived, and those they act on are able to strategically develop plans for success. While this may present challenges in applying insights from r/K selection theory to strategic environmental communication, we hope to manage these challenges by orienting this chapter towards practice as a means of helping communicators take advantage of environments with low uncertainty, reducing uncertainty when possible, and developing strategies to cope with low uncertainty when not possible. While reflexivity creates some issues in application, it also ultimately makes this work impactful because it allows this work to guide conscious communication strategy.

Another potential limitation is the question of resources and scale. The theory of r/K selection is premised on the degree of resource investment, which in biological studies often refers to energy or time investment. While time investment may work similarly for prospects across disciplines (offspring vs. communication strategies), energy does not have a clear equivalent in the field of strategic environmental communication. The closest equivalent may be capital, which is a requirement for the development and implementation of communication strategies, which we have discussed throughout this chapter. That said, capital and energy do not function identically, especially when it comes to scale. While an organization can greatly expand its communication plans when it has access to more capital, an organism cannot generally make use of greater energy levels beyond a certain point. If an animal is no longer in danger of starvation for the foreseeable future, consuming more food will not necessarily increase its chances of leaving successful offspring. One of the biggest issues in application on this axis is that well-funded strategic communication entities may have less cause for concern about the costs of taking a varied communication approach in addition to a highly specialized approach. In this case, the insights from this work may be less applicable to these communicators unless they are interested in pursuing a parsimonious approach to communication strategy.

In addition to limitations to the technical applications of r/K selection theory to the study of communication environments, we must also discuss limitations in how this theory is used by communication practitioners. We should be very careful in noting that while this application is presented as a holistic theory of environmental uncertainty, communication practitioners should not use it to erase the nuance of different features of their communication environments. While the composition of stakeholder groups of environmental organizations and the media employed by those organizations may both be thought of in terms of environmental uncertainty, they are distinct features of an organization's communication environment. An organization can strategically position itself in different ways by adjusting many elements that comprise its communication environment, and these elements should not be treated as equivalent simply because they can be seen through a lens of environmental uncertainty. Rather than using the insights from this chapter to flatten the nuance of their communicative strategy, we hope that the communication practitioners who read this chapter may be emboldened to think about how they can plan their strategy from a high level by considering the features of their whole communicative environment.

Finally, it cannot be emphasized enough that this chapter is the first attempt at applying the insights of r/K selection theory to the field of strategic environmental communication. Thus, the ideas presented are highly exploratory in nature, and more explication and research are required before the claims presented in this chapter have a solid foundation in strategic communication research. While the theory of r/K selection is fairly well established as a means of understanding the pressure that environmental uncertainty places on biological species (MacArthur and Wilson 2001), we have already discussed some issues in application to strategic environmental communication. With all likelihood, additional issues in the application that have yet to be considered will only be uncovered as more research is conducted. As it stands, we can only make claims that are guided by the principles grounded in biological science and make an argument that environmental uncertainty is a relevant concept that warrants further investigation in the field of strategic environmental communication.

10.6. Conclusion

Now more than ever, scholarly research guiding the effective use of strategic communication for key environmental goals is needed. This chapter draws from studies in biology to propose an environmental communication strategy that is guided by an evidence-based understanding of the natural world. Inspired by r/K selection theory, this chapter presents insights for strategic environmental communicators considering environmental uncertainty as a key factor in their strategic planning. For communicators operating in relatively stable environments, characterized by homogenous groups of key stakeholders who can be reached via a small number of media channels, we argue that they can benefit from a highly specialized approach to communication strategy. Communicators for whom this is not the case may attempt uncertainty reduction to make their environment more favorable to specialized approaches, which often entails learning more about their key groups of stakeholders and possible media channels. Uncertainty reduction will not always be effective, however, and there will be many occasions when a communicator's environment will be characterized by uncertainty deriving from outside of their control. In these cases, we reflect back on the value of highly distributed strategies that cope with uncertain environments in the field of r/K selection. At the end of this exploratory interdisciplinary work, we close on the importance of considering environmental uncertainty not only in communication practice but also in research that tests its specific application in the field of strategic environmental communication. We hope this chapter serves as the impetus for research that tests these ideas, and we hope this course of study ultimately provides value to environmental communicators working to rise to environmental challenges at such a critical time.

10.7. References

Atkinson, L. (2012). Buying in to social change: How private consumption choices engender concern for the collective. *The Annals of the American Academy of Political and Social Science*, 644, 191–206.

Botan, C.H. and Taylor, M. (2005). The role of trust in channels of strategic communication for building civil society. *Journal of Communication*, 55(4), 685–702. doi: 10.1111/j.1460-2466.2005.tb03017.x.

Brunnander, B. (2007). What is natural selection? *Biology & Philosophy*, 22(2), 231–246. doi: 10.1007/s10539-005-9008-4.

Cassill, D.L. (2019). Extending r/K selection with a maternal risk-management model that classifies animal species into divergent natural selection categories. *Scientific Reports*, 9(1), 1. doi: 10.1038/s41598-019-42562-7.

Clean Creatives (n.d.). The future of creativity is clean [Online]. Available at: https://cleancreatives.org.

Crane, A. and Ruebottom, T. (2011). Stakeholder theory and social identity: Rethinking stakeholder identification. *Journal of Business Ethics*, 102(1), 77–87. doi: 10.1007/s10551-011-1191-4.

Dalrymple, K.E., Shaw, B.R., Brossard, D. (2013). Following the leader: Using opinion leaders in environmental strategic communication. *Society & Natural Resources*, 26(12), 1438–1453. doi: 10.1080/08941920.2013.820812.

Darwin, C. (1859). *The Origin of Species by Means of Natural Selection or the Preservation of Favoured Races in the Struggle for Life*. John Murray, London.

Darwin, C. (1909). *The Voyage of the Beagle*. P.F. Collier, New York.

Eccles, R.G., Lee, L.-E., Stroehle, J.C. (2020). The social origins of ESG: An analysis of innovest and KLD. *Organization & Environment*, 33(4), 575–596. doi: 10.1177/1086026619888994.

Fernández-Moreno, M.A., Farr, C.L., Kaguni, L.S., Garesse, R. (2007). Drosophila melanogaster as a model system to study mitochondrial biology. *Methods in Molecular Biology*, 372, 33–49. doi: 10.1007/978-1-59745-365-3_3.

Friedman, A.L. and Miles, S. (2002). Developing stakeholder theory. *Journal of Management Studies*, 39(1), 1–21. doi: 10.1111/1467-6486.00280.

Fuller, D. (1999). *Sustainable Marketing: Managerial-Ecological Issues*. SAGE Publications, Thousand Oaks, CA. doi: 10.4135/9781452220611.

García, C. (2012). Using strategic communication for nation-building in contemporary Spain: The Basque case. *International Journal of Strategic Communication*, 6(3), 212–231. doi: 10.1080/1553118X.2012.678523.

Gordon, K. (2004). On survival of the fittest and other corporate myths. *World Futures*, 60(8), 617–628. doi: 10.1080/026040290504512.

Hallahan, K., Holtzhausen, D., van Ruler, B., Verčič, D., Sriramesh, K. (2007). Defining strategic communication. *International Journal of Strategic Communication*, 1(1), 3–35. doi: 10.1080/15531180701285244.

Hanus, M.D. and Fox, J. (2017). Source customization reduces psychological reactance to a persuasive message via user control and identity perceptions. *Journal of Interactive Advertising*, 17(1), 1–12. doi: 10.1080/15252019.2017.1287023.

Heide, M., von Platen, S., Simonsson, C., Falkheimer, J. (2018). Expanding the scope of strategic communication: Towards a holistic understanding of organizational complexity. *International Journal of Strategic Communication*, 12(4), 452–468. doi: 10.1080/1553118X.2018.1456434.

Hildebrandt, T., Drews, B., Gaeth, A.P., Goeritz, F., Hermes, R., Schmitt, D., Gray, C., Rich, P., Streich, W.J., Short, R.V. et al. (2007). Foetal age determination and development in elephants. *Proceedings of the Royal Society B. Biological Sciences*, 274(1608), 323–331. doi: 10.1098/rspb.2006.3738.

Hirsh, J.B., Kang, S.K., Bodenhausen, G.V. (2012). Personalized persuasion: Tailoring persuasive appeals to recipients' personality traits. *Psychological Science*, 23(6), 578–581. doi: 10.1177/0956797611436349.

James, S.P. (2014). "Nothing Truly Wild is Unclean": Muir, misanthropy, and the aesthetics of dirt. *Environmental Ethics*, 36(3). doi: 10.5840/enviroethics201436335.

Jørgensen, S.E. and Mejer, H. (1979). A holistic approach to ecological modelling. *Ecological Modelling*, 7(3), 169–189. doi: 10.1016/0304-3800(79)90068-1.

Kimmerer, R. (2013). *Braiding Sweetgrass: Indigenous Wisdom, Scientific Knowledge and the Teachings of Plants*. Milkweed Editions, Minneapolis, MN.

Liang, Y., Kee, K.F., Henderson, L.K. (2018). Towards an integrated model of strategic environmental communication: Advancing theories of reactance and planned behavior in a water conservation context. *Journal of Applied Communication Research*, 46(2), 135–154. doi: 10.1080/00909882.2018.1437924.

MacArthur, R.H. and Wilson, E.O. (2001). *The Theory of Island Biogeography*. Princeton University Press, Princeton, NJ.

Moyer-Gusé, E., Tchernev, J.M., Walther-Martin, W. (2019). The persuasiveness of a humorous environmental narrative combined with an explicit persuasive appeal. *Science Communication*, 41(4), 422–441. doi: 10.1177/1075547019862553.

Muir, J. (1897). The American forests. *The Atlantic* [Online]. Available at: https://www.theatlantic.com/magazine/archive/1897/08/the-american-forests/305017/.

Muir, J. (1901). *Our National Parks*. Houghton Mifflin and Company, Boston, New York, [Online]. Available at: http://archive.org/details/ournationalparks00mu.

Muir, J. (1920). *The Yosemite*. Century Company, New York.

Murdock, E.G. (2021). Conserving dispossession? A genealogical account of the colonial roots of western conservation. *Ethics, Policy and Environment*, 24(3), 235–249. doi: 10.1080/21550085.2021.2002625.

Pew (2023). Social media fact sheet. *Pew Research Center: Internet, Science & Tech* [Online]. Available at: https://www.pewresearch.org/internet/fact-sheet/social-media/.

Rathnayake, C. and Winter, J.S. (2018). Carrying forward the uses and Grats 2.0 Agenda: An affordance-driven measure of social media uses and gratifications. *Journal of Broadcasting and Electronic Media*, 62(3), 371–389. doi: 10.1080/08838151.2018.1451861.

Rogers, R. (2020). Deplatforming: Following extreme Internet celebrities to Telegram and alternative social media. *European Journal of Communication*, 35(3), 213–229. doi: 10.1177/0267323120922066.

Rose, J., Pitt, A., Verbos, R., Weller, L. (2022). Incorporating movements for racial justice into planning and management of U.S. National Parks. *Journal of Park & Recreation Administration*, 40(1), 44–60. doi: 10.18666/JPRA-2021-11005.

Rosen, P. (2023). Remember when NFTs sold for millions of dollars? 95% of the digital collectibles may now be worthless. *Markets Insider* [Online]. Available at: https://markets.businessinsider.com/news/currencies/nft-market-crypto-digital-assets-investors-messari-mainnet-currency-tokens-2023-9.

Soedarsono, D.K., Mohamad, B., Akanmu, M.D., Putri, I.P. (2020). Political leaders and followers' attitudes: Twitter as a tool for political communication. *Journal of Critical Reviews*, 7(8), 1245–1252.

Sundar, S.S. (2008). *The MAIN Model: A Heuristic Approach to Understanding Technology Effects on Credibility*. MacArthur Foundation Digital Media and Learning Initiative, Cambridge.

Sundar, S.S., Qian, X., Dou, X. (2019). The role of technology in online persuasion: A MAIN model perspective. In *Advertising Theory*, Rodgers, S. and Thorson, E. (eds). Routledge, New York.

Weinhold, B. (2006). Epigenetics: The science of change. *Environmental Health Perspectives*, 114(3), A160–A167.

Werder, K.P., Nothhaft, H., Verčič, D., Zerfass, A. (2018). Strategic communication as an emerging interdisciplinary paradigm. *International Journal of Strategic Communication*, 12(4), 333–351. doi: 10.1080/1553118X.2018.1494181.

Zahn, M. (n.d.). What is ESG investing and why are some Republicans criticizing it? ABC News [Online]. Available at: https://abcnews.go.com/Business/esg-investing-republicans-criticizing/story?id=97035891.

Zerfass, A., Verčič, D., Nothhaft, H., Werder, K.P. (2018). Strategic communication: Defining the field and its contribution to research and practice. *International Journal of Strategic Communication*, 12(4), 487–505. doi: 10.1080/1553118X.2018.1493485.

11

Biodiversity in Discourse Surrounding CSR: What Narrative Framework is Needed to Prevent "Biodiversity-washing"

There are two "mega-themes" which are rapidly rising to the top of corporate agendas and shifting the way in which businesses are organized: the climate crisis and the biological crisis. While the former has been on the agenda of many businesses over the past twenty years, through the thrust of networks and both soft and hard laws which have manifested in the development of referential frameworks, labels, new processes and ad hoc systems, the latter has only recently crossed the horizon of the same businesses, seemingly through the back door.

We can describe them as "mega" in the sense that these two topics, which bring an added layer of complexity for businesses, have the quality of being universal in terms of their consequences, transcending national boundaries, while also having an impact at local level, and affecting businesses differently depending on the sectors in which they operate.

The two issues were merged on the international institutional agenda several years ago, with the first IPCC1-IUCN2 working group having delivered its conclusions in June 2021. Therefore, the scientific agenda is not yet set in stone by consensus; however, even at this stage, businesses are being asked to make efforts in these areas, and taking account of biodiversity both in their accounting and in their strategy.

Chapter written by Céline PASCUAL ESPUNY and Catherine LONEUX.

In terms of communication, twenty years on, businesses need to avoid the same kind of mistakes as happened with so-called greenwashing, while seeming to retain the same stance: when integrating biodiversity into the climate narrative on the business agenda, how can they avoid what might be termed biodiversity-washing?

Furthermore, other subsidiary research questions arise: from what angle should we approach biodiversity in the world of business? How can we integrate biodiversity into a narrative that already hinges on CSR (Corporate Social Responsibility) and sustainable development? When and why should we communicate? What is the role of established networks, how are they organized and what role should they play in shaping a new kind of communication that is not discredited?

After presenting a state of the art, which allows us to appreciate the communicational context in which this new mega-theme has emerged for businesses, we will analyze this hybridization of discourse in a context that is marked by greenwashing. We offer a critical and prospective interpretation of this social and communicational reality, looking at both the emergence of public problems for corporate communications and the issues of discourse and narrative strategies relating to communication on CSR. Finally, in the third section, we identify the main feature of these narrative mega themes and the risks of biodiversity washing that occur in relation to this new emerging narrative through a study of publications and reports published on the platform of the "*Initiative française pour les entreprises et la biodiversité*"[1], and those produced in the working group "Entreprises et Biodiversité" (Business and Biodiversity), set up by the Comité français de l'UICN in June 2009, which has 66 members (IUCN personnel, their private-sector partners and experts)[2].

11.1. Climate and biodiversity: the recent convergence of two parallel narratives

Although driven by two different scientific worlds (climatology and geoscience, on the one hand; life sciences and environmental sciences, on the other hand), the alarm bells regarding climate change and biodiversity loss have followed paths that were parallel until very recently.

1 See: https://entreprises-biodiversite.fr/.

2 See: https://uicn.fr/entreprises-et-biodiversite/.

"For a long time, we have viewed the climate and biodiversity as being two different things", Pamela McElwee stated in a press conference; McElwee lectures in human ecology at Rutgers University (United States):

> Politicians have also followed the same path [...]. The importance attached to the climate has gradually increased – I believe because we are feeling the effects, although the bushfires in Australia may seem far away. Nevertheless, the two things are very closely linked[3].

Born of two different seminal texts, but from the same conference, the UNCED[4], also known as the Earth Summit, in Rio de Janeiro in 1992, the Framework Convention on Climate Change and the Convention on Biological Diversity (CBD) each established their own international institutional agenda. Historically, each of these mega-topics came into the public eye in a different way. The matter of biodiversity began receiving attention from the 1780s, but only among biologists and naturalists, until the 1980s (Raffin 2007)[5], whereas climate change was immediately catapulted into the public eye.

In the public arena, the issue of the climate dominates debate; every year, there are *Conferences of the parties* (COPs), which receive widespread media attention[6], giving rise to multiple offshoots and institutional programs

3 See: https://www.novethic.fr/actualite/environnement/biodiversite/isr-rse.

4 United Nations Conference on Environment and Development.

5 According to Raffin (2007), awareness of the fragility and protection of "the Wild" began with the Conquest of the American West and the colonization of wild territories. It became concrete in 1860, under the impetus of John Muir, by the establishment of sanctuaries in remarkable spaces, which would later become National Parks. The management of the parks and restrictions on fishing and hunting gave rise to the need for the first International Conference for the Protection of Nature, in Bern, in 1913. This led to the founding of an Advisory Committee for the International Protection of Nature – the forerunner of today's IUCN. Biodiversity loss has been a matter of concern for certain biologists since the 1960s. Scientific reports raising the alarm, and in particular, Rachel Carson's 1962 book *Silent Spring*, and Jean Dorst's, *Avant que nature meure* (Before Nature Dies), in 1965, finally catalyzed the public into action.

6 To date, 29 climate change COPs have been held, some of them receiving a huge amount of media attention: for example, COP 3 in Kyoto, COP 15 in Copenhagen, COP 18 in Doha, COP 21 in Paris, and COP 26 in Glasgow.

to lead the fight against climate change, involving all actors: scientists, businesses, institutions and civilian society.

On the contrary, biodiversity COPs take place every two years, and there has never been one that has received similar attention from the world's media. However, the two phenomena are on a meta level: they are linked and both as disastrous as one another (Parrenin and Vargas 2020). Is this linked to the weight of the actors, networks and the implications?

11.1.1. *How can we explain the so-different dynamics of emergence between the issues of climate and biodiversity?*

The Earth Summit in 1992, in Rio de Janeiro, birthed the two Framework Conventions. It also established an open logic of participation of civil society and businesses in relation to the major key questions which had, up until that point, only been addressed by experts, scientists and diplomats in the confines of the United Nations (UN).

There were only five years between the ratification of the Brundtland Report, Our Common Future, which set the definition of sustainable development in the introduction to Chapter 2 (CMED 1987), and the Earth Summit. A number of bodies were set up, charged with implementing what had been stipulated in Stockholm (the first Earth Summit, 1972), in the directives of the United Nations Environment Program (UNEP) and Nairobi (the second Earth Summit, 1982). The IPCC was set up in 1988, as was the Centre for Our Common Future in Geneva; 1989 saw the founding of the Stockholm Environment Institute and 1990 saw the establishment of the Global Environment Facility.

The Global Forum marked the intervention of non-governmental organizations (NGOs), and also the world of business, in the debates, notably about ethical business standards (Loneux 2010).

In order to organize in regard to this new issue, take action and apply the prerequisites of the Rio Conventions, businesses created new networks and organizations to fight climate change (World Business Council in Sustainable Development, Global Reporting Initiative (GRI) in 1997; the Ceres, Coalition for Environmentally Responsible Economies, set up in 1989; and the Global Compact, in 2000, driven forward by the UN).

Playing a part in working groups, producing reports and indicators (carbon balance, socially responsible investing, CSR reports) or standards (ISO 9001, 14000, 26000), businesses have, little by little, been endowed with processes and toolkits, allowing them, at their own level, to combat climate change. Those businesses which engage with these goals carry out assessments, change direction and produce more sustainably. In this regard, the deontological codes produced by the International Chamber of Commerce in Paris offer an impressive arsenal in terms of environmental "best practices" (Loneux 2016).

In parallel, there was little discussion of the topic of biodiversity. The equivalent of the IPCC, the IPBES (Intergovernmental Science-Policy Platform on Biodiversity and Ecosystem Services), was slow to be implemented. Placed under the aegis of the UNEP, the United Nations Development Programme (UNDP), the United Nations Educational, Scientific and Cultural Organization (UNESCO) and the Food and Agriculture Organization (FAO) of the United Nations, the IPBES was officially established on April 21, 2012, by 94 governments. It held its first meeting in early 2013.

There are not, strictly speaking, networks devoted to biodiversity in businesses as there are for climate issues: in the majority of cases, it is either international bodies (such as the IUCN and IPBES) which, at the initiative of a few large groups, host a working group on Business and Biodiversity; or else, the networks set up to address the climate crisis which, ultimately, also work on the same logics and the same organizational bases, on problems relating to biodiversity.

Thus, in fact, we see a reappropriation and an inclusion of the narrative matrix on biodiversity within the climate matrix; however, the issues and their impacts are indeed different.

Thus, what can be done to avoid repeating the mistakes of greenwashing committed by many businesses, through lack of knowledge but also sometimes for reasons of ease, when they first began talking about their corporate social and environmental responsibility? How can we move toward CSR which is truly "progressive" (in the sense of progress in environmental ethics)?

11.1.2. *Two profiles for two different narrative frameworks*

In summary, while the situation is, of course, not the same for businesses today trying to integrate topics taking account of biodiversity into their discourse, the pitfall to be avoided has the same profile: the knowledge and skills in terms of biodiversity are so tenuous that it is tricky to avoid falling into hollow, generalist discourse.

Today, speaking of biodiversity in a corporate environment is a delicate balancing act, in view of the negative balance created by long years of accusations of greenwashing and pressure from associations.

Only a few multinationals, who have been able to anticipate and prepare for the arrival of this topic in view of the sector in which they operate, or their engagement with a CSR approach, are leading the way, and have been able to develop original programs and integrate indicators allowing them to take account of biodiversity in their CSR reports. For others, who have struggled to integrate so-called sustainable communication, it is as complex to integrate environmental concerns into their discourse as it was to integrate the fight against climate change. The difference is that now, they have no room for error: they are being closely watched by environmental associations and also by civil society, which has acquired skills and knowledge that almost parallel those of scientists.

Indeed, biodiversity has a participative, familiar and incarnate profile, which the fight against climate change did not have: natural sciences were among the very first disciplines to make use of strategies such as open science, citizen sciences or science shops, calling on the services of amateur naturalists, who had no hesitation in engaging with the open data systems that were made available to them:

> In questioning the objectives of the body of specialized researchers, citizens have built up skills relating to the social use that they can make of scientific research. This has led to the emergence of a new type of relationship between professional researchers and citizens and their structures (private structures such as associations, or public ones such as municipal authorities) – which, in turn, has had an impact on researchers and their laboratories (Le Crosnier et al. 2013).

Consequently, over the last decade or so, civil society has gained great familiarity, firstly, with the issues relating to biodiversity, and secondly, with the associations and networks of actors whose professional goal or commitment is to the protection of species.

In summary, the recent but powerful convergence of these two mega-themes from an international institutional agenda to a business agenda could foreshadow a possible and renewed slide into biodiversity-washing: the associative and societal knowledge of the topic comes first, and it is more detailed and more anchored socially. Once again, it is a question of urgency, although "facts are uncertain, values are controversial, stakes are important and decisions are urgent" (Funtowicz and Ravetz 1991, 1993). In communicational terms, the germs of a potential crisis, or at least, renewed disenchantment and accusations of greenwashing, are there: uncertainty, urgency and conflicts of values that are not seen by companies not mature enough to measure public sensitivity about matters of biodiversity mean that when they speak about the subject, it could give rise to a crisis. How are we to integrate discourse about biodiversity, where the issues are linked, and of equivalent importance, to the pre-existing discourse about climate change? How can we ensure that biodiversity is given its proper place in corporate communications?

11.2. Biodiversity: bringing about a shift in CSR communication and avoiding the risk of greenwashing

Biodiversity is a field of communication that is equally broad and ample as that of climate change, in terms of the problems that it raises, but its qualities are different. It is less ethereal, more tangible, easier to grasp for a business and also for the general public.

Speaking of biodiversity involves speaking of plant and animal species, some of which are particularly iconic – notably mammals, in the animal kingdom; fruits and vegetables, for the plant kingdom; and also, some highly symbolic insects. Thus, it has been easy for NGOs to mobilize large numbers of sympathizers to change practices in businesses by creating reputational crises around the symbolic defense of the habitat of big tigers or orangutans (e.g. consider Greenpeace's campaigns against Kit Kat/Nestlé). The public is aware and is concerned about the extinction or fragility of certain iconic species (pandas in the case of the WWF, seals, coral reefs, whales, polar bears, koalas, turtles, etc.).

Certain businesses – particularly major multinationals engaged in CSR – were soon called upon, or threatened, by NGOs for their practices and those of their subcontractors which posed a direct threat to biodiversity.

In order to better understand the issues raised in terms of communication, at the intersection between those relating to organizational communication and environmental communication, we start with the overall understanding in the context in which CSR communication developed, before discussing the question of building public problems into corporate discourse. We will also examine how earlier practices of greenwashing have connotations for the exercise of integrating the themes of biodiversity into current CSR communication, making it more difficult. Finally, we will look at the role of networks in the communications of the future.

11.2.1. *A global discursive foundation: CSR communication*

Understanding communication whose purpose is to express businesses' CSR commitment, first and foremost, involves understanding the social evolution that has taken place in terms of the meaning attached to the concept of development and understanding how organizations have made use of the concept. As Tremblay (2007, p. 10) points out, development is a concept loaded with ideologies, and the meaning that is currently attached to the word "development", in the expression "sustainable development", conditions and determines the self-appointed role and actions of organizations in the vast movement of standardization and commitment that is ongoing.

Just as the concept of sustainable development has been shown to be fuzzy and plastic (Pascual Espuny 2007), that of CSR is also ambiguous. However, as Lapointe (2007, p. 67) indicates, its definition is undoubtedly vague, but it highlights that social responsibility lies in going above and beyond what is mandated by law and regulations, and that businesses are not responsible to their shareholders alone:

> Normative schemes such as soft laws are emerging around CSR, driven by communication, leading to a constant stream of sociocognitive, individual and collective adjustments. Ideologies circulating within major organizations are seeping into the management: rationalization of knowledge by "deontological" values, dissemination of "ethical" knowledge, production of a

> specific framework to establish exchanges between corporations and society at large, standardized and institutionalized logics of action, and the creation of sustainable collectives (Loneux 2016, p. 2).

The idea of CSR has given rise to new stances and attitudes on the part of businesses, expressed through discourse and communication channels aimed not only at shareholders, and in this context, a number of powerful themes, such as ethics and transparency, have come to light and have greatly impacted the communication of all these actors. As Allard-Huver (2015, p. 35) states:

> Sustainable development is a very particular rhetorical environment, where transparency has a specific relationship with environmental objects: rendering the invisible visible; 'demonstrative' discourse, so to speak (Aïm 2006; Jeanneret 2010).

In this initial framework, we see the construction of the first attempt by companies engaged in CSR to express their vision and commitment to the fight against climate change. However, in the overall context of narrative creation around a new topic that is not well understood but is attractive, particularly plastic and universal, they switched to a practice that was soon declaimed as greenwashing.

11.2.2. *Greenwashing: a heavy burden from past wrongdoing*

Greenwashing referred to the fact of misleading consumers in relation to a company's environmental practices ("firm-level greenwashing") or the environmental benefits of a product or service ("product-level greenwashing") (TerraChoice 2010)[7].

It is easy to understand greenwashing as a distortion between the image that a business projects and communicates, and its reality, and also a defect in the consistency between its actions and its communications (Pascual Espuny 2015). According to Delmas and Cuerel Burbano (2011), greenwashing is at the intersection between two corporate attitudes: poor environmental performance and positive communication about it. The effect of greenwashing is to undermine the credibility of environmental discourse from all angles, whether justified or not (Vos 2014).

7 See: http://sinsofgreenwashing.com/findings/greenwashing-report-2010/index.html.

In its wake, the denounced practice of greenwashing led to a certain number of communicational levers that are symbolic of CSR: thus, the use of green coloring (Biros 2014), which is widely used to refer implicitly to sustainable development, and the use of the adjective "sustainable" or self-assigned labels, are counterproductive, and at best lead to disinterest, or to suspicion (Parguel et al. 2011).

Furthermore, it leads to a profound crisis for corporate communication, placing them in a situation of paradoxical injunction: they are supposed to have ethical discourse and be assessed on the basis of their environmental performance; nevertheless, they are not taken into consideration when they communicate on that subject (Coppolecchia 2010).

It is a particularly tricky task to communicate about the climate crisis in this context: businesses have had to learn to compose with silence, or to only express their convictions in terms of what is factual, what is measurable, in their area of expertise. With the phenomenon of discourse restriction, which notably was imposed at the level of publicity self-regulation, communicational mutism was established about the business' commitment, at its own level, to the fight against climate change.

Greenwashing created a complicated precedent for communication, although indeed communication is required by the market, by stakeholder and by society as a whole. Indeed, companies are forced to confront the question of their reputation, in terms of their capacity to put their messages across through their environmental communication while avoiding greenwashing.

11.2.3. *Marrying urgency with narrative frameworks*

In this context, how are we to integrate a new issue that is so broad and complicated as that stemming from biodiversity loss?

This broadening field of discourse, relating to a serious, global, complex and thorny issue, can be compared, in information and communication sciences, to theoretical studies on social problems. According to Neveu (1999 p. 42), "the so-called 'social problem' is none other than the transformation of some social reality into an issue for public debate and/or State intervention".

Delforce and Noyer (1999, p. 24) point to a discursive and social phenomenon characterized by tension, indicating specific social discourse, "provisionally stabilized [...] relating to types of occurrences/events or to themes, resulting from the operation of this global social discourse while businesses allow it and nourish it".

These authors point to the importance of the discursive dimension in the introduction and understanding of new problems in the public domain. They note processes of definition and transaction that are established, with variable issues. They also point out the importance of discursive communities, which are occasionally established to defend or support a definition of the problem.

In this context, how are we to understand the emergence and, particularly, the circulation of topics relating to biodiversity in corporate discourse? As we have seen, it is a matter of integrating a topic that has matured in the shadow of that of climate change and which, while it is not new, is nevertheless still fuzzy in the way in which it is integrated into corporate practices.

In terms of communication, what are the issues at hand? The dominant matrix is that of climate change. This comes before anything else and has impacts on the issue of biodiversity. What can companies say about the extent and range of global biodiversity loss? What legitimacy do they have to speak, notably for businesses which do not have an immediate impact on biodiversity? Should they, therefore, stay silent and not engage with this emerging topic, which is more worrying but also more familiar to us than that of climate change?

11.3. An immature narrative, calqued from that of climate: analysis of the initial relations linking climate and biodiversity

In this final section, we will compare our statements to discourse analysis, in order to better describe this hybridization and the routes it takes. Topics relating to biodiversity have only very recently been built into the political agenda, so we will base our discussions on the work of pre-existing working groups, who have worked hand in hand with international organizations in the field, such as the IUCN and the IPBES.

We will analyze the content of 15 reports and publications cowritten by experts from the IPBES and the IUCN and delegates appointed by businesses to contribute to the drafting of those documents. We will look at the only two public–private business/biodiversity initiatives currently in France, where our research is focused. These are two different entities. One is part of the IUCN – the group *Entreprises et Biodiversité* (Business and Biodiversity). The other is the platform of the *Initiative française pour les entreprises et la biodiversité* (French Business and Biodiversity Initiative).

In terms of methodology, we opted to perform a case study. The research question we attempted to answer was such a recent issue that we cannot perform an in-depth study of an extensive corpus. Instead, the corpus is restricted to a number of actors, representing their institution in working groups and in defining a new topic for businesses. In this context, a field study is the most appropriate way to proceed.

The second choice made relates to discourse analysis of reports and publications produced, because it is the interdiscursive framework that is of interest to us at this stage of our research. We seek to understand how topics relating to biodiversity are perceived and communicated, and in what way they are presented and translated into the perception of the businesses in question. In addition, this research aimed to understand the place given to networks, as emitters and sources of credibility for this discourse, given that greenwashing practices have destabilized businesses as a credible source of information for broad discourse about sustainable development.

Thus, in June 2009, the French chapter of the IUCN set up a working group, *Entreprises et Biodiversité*, bringing together the members of the IUCN, their private-sector partners (EDF, ENGIE, Ciments Calcia-GSM, Eqiom and Veolia) and 60 experts. The group is currently chaired by Denis Leca (Storengy), who took over the helm from Thierry Hauchard (GSM 2011–2013) and Claire Varret (EDF 2014–2017). The working group *Entreprises et Biodiversité* is an arena for discussions and proposals to generate involvement from businesses in relation to biodiversity. It meets two to four times per year[8]. The working group has carried out numerous projects, whose conclusions and publications are available for download from the website:

– raising awareness and motivating the companies' employees to act in favor of biodiversity;

8 See: https://uicn.fr/entreprises-et-biodiversite/.

– managing invasive exotic species;

– biodiversity reporting and indicators used by the businesses;

– ecological offset;

– private-sector biodiversity training;

– biodiversity: an asset for your corporate sites.

The platform of the French Business and Biodiversity Initiative is a much further-reaching and more active initiative than the IUCN's *Entreprises et Biodiversité* working group. The Initiative's website is exhaustive, informative, sourced and shows numerous ideas and resources that businesses have used to improve their impact in terms of biodiversity. Set up in late 2013, the platform is defined as "a gathering point open to any and all organizations wishing to feed into exchanges about the issues of safeguarding biodiversity and ecosystem function"[9].

The platform provides visitors with numerous documents, annual reports (from 2015, 2016 and 2018), a library of documentary resources, guides and publications, reports and case studies. The website is well populated and professionally presented.

11.3.1. *Hesitant communication*

The 30-odd technical documents, reports, videos, situation reports, case studies, resource documents, press releases and brochures that we have analyzed, across the websites of *Entreprises et Biodiversité* at IUCN France and the platform of the *Initiative française pour les entreprises et la biodiversité*, show communication that is tacit, which is particularly reserved and hesitant, but also contains hidden gems which occasionally come to light.

Semiological and semantic analysis reveals points in common: firstly, and unsurprisingly, the iconography, which is blue and white, is based on a profusion of photos of animals and plants, in the form of inserts, sometimes with a label and sometimes without, which provide occasional illustration for all the texts, with no obvious connection between them. For example, in the illustrated literature, we find multiple pictures of small animals, highlighted and highly colorful, in wordy texts on a plain white background. Whatever

9 See: https://entreprises-biodiversite.fr/la-plateforme/a-propos/.

the medium, the graphic effort is basic, both in the drop capitals and the typefaces used. There is no image blending or perspective, and similarly, no attempt at subheadings. The whole reveals a dire situation.

11.3.2. *Finding one's place*

The second general observation relates to the tone adopted and to each actor's assigned role: while the two media analyzed here result from fifteen years of work, the content analysis shows a sometimes-clumsy attempt to find common ground and shared culture. For instance, an organization like the IUCN seeks to encode its documents in a collegial manner, highlighting the contributors to the working group, but it cannot resist the temptation of presenting itself as the main source, oscillating between a stance of a knowledgeable player and an expert or sometimes a judge. The guides and factsheets published and available on the site – "deep-sea mining", "palm oil biodiversity" and "marine plastics" – are representative of this oscillation between efforts to make knowledge more broadly accessible and to sit in judgment. The reports such as the "*Guide sur les espèces exotiques envahissantes*" (Guide to Invasive Exotic Species) or "*Reporting biodiversité*" (Biodiversity Report) are explicit here: over half the pages are devoted to explaining each party's roles and also their limitations, prudent and legal reminders, or appropriation of codes. Thus, we were surprised to see an explanation of what annual reports, financial reports and CSR reports are, in documents for a readership whose very function is to write such reports in businesses. Whilst we see this quest for well-foundedness and legitimacy in the environmental organizations, the businesses' stance is also ambiguous: only a handful of companies have taken part in these working groups, and they often present themselves as good students in documents designed "for others".

This introductory passage summarizes this position clearly:

> Businesses have a major role to play in reducing their biodiversity impact, but also in restoring and highlighting the value of biodiversity and ecosystems. The French chapter of the IUCN has been working for many years to improve the involvement of businesses in favor of biodiversity[10].

10 See: https://uicn.fr/entreprises-et-biodiversite/.

The confusion is also visible in the real-world actions: many cliches concerning businesses are present in the reports – notably in terms of the benefits that they can derive from taking account of biodiversity. Without explanation, arguments of authority are served: "improving the image and reputation of the business [...] improving employees' wellbeing". The confusion is also visible in the encoding of the message to interest businesses: the choice has been made to speak of "services rendered", quantified and valued financially. However, the method is not given, and it is indicated, in the preamble, that "preserving biodiversity is an economic issue, but it is not a matter of putting a price on Nature", although in the next paragraph, this is what is attempted.

In the texts, however, we see a number of concepts which emerge and which are intended to be stabilized and developed in years to come – the fruits of clever encoding between, firstly, the pre-established indicators and logic for the struggle against climate change, and secondly, the logic specific to biological considerations: hence the subdivision of the notion of "natural capital" and "environmental performance" which "is subject to non-financial rating". We also see the emergence of a breakdown of the services rendered: "supply service", "regulation services", "cultural services", "support services" and "housing and support service".

Finally, it is primarily in the logic of networking adopted by the platform that we find the richest communicational avenues for the future: the platform offers a logic of networking and valuing of these members, which presents that of the Global Compact. Positioned as a "gateway", it is read as a hybrid extranet/Internet site, dedicated to businesses considered as partners, in a form of communication that is primarily horizontal and participative. The approach is also incentive: in addition to the resource sections, there are "case studies" and "eco-lex" – a link to a different platform that compiles all global environmental applying to businesses – "Feedback" and "Networks" sections. This stance is much more complex and much more dynamic, in the role of a mediator and a bridge between multiple targeted initiatives relating to biodiversity (forest defenses, coastal defenses, construction of blue and green frameworks, work on a mechanism of eco-systemic offset, etc.) which have flourished among private businesses. The logic of networking and positioning here engages a humbler stance because it is more fallible and more reflective, and also more dynamic. The compilation of case studies is up to date and vibrant and can also be seen in the form of short testimonial videos. It therefore becomes accessible.

11.4. Conclusion

By way of conclusion, it seems that corporate communication on biodiversity is rapidly gaining its place in global corporate discourse, by way of a relatively hesitant rhetorical framework in the words and concepts, but increasingly mature and fine in its circulation between sources. For all of the players involved, it is a matter of integrating a mega-theme into an existing, pre-established framework, codified and stabilized. While for the communicators, communicating about biodiversity is a high-wire act, it seems nevertheless that it is beginning to find its position.

Biodiversity is an emerging topic, which is likely to grow very rapidly in corporate discourse. Shifting pre-existing mainstream CSR discourse towards a biological component without falling into the cliches of "bambification" (Milstein 2011) for environmental actors, to remain credible while taking account of neglected aspects and embracing an eco-systemic conception of the environment in the face of productive and competitive discourse, avoiding the major pitfall for the announcers and if the key were in good management of communication networks dedicated to hybrid climate and biodiversity topics?

11.5. References

Aïm, O. (2006) La transparence rendue visible. *Médiations informatiques de l'écriture, Communication et langages*, 147, 31-45.

Allard-Huver, F. (2015). La transparence, de pierre angulaire à pierre d'achoppement des controverses médiatiques. In *La communication transparente, l'impératif de transparence dans les discours des organisations*, Catellani, A. (ed.). Presses Universitaires de Louvain, Louvain la Neuve, 31–47.

Biros, C. (2014). Les couleurs du discours environnemental. *Mots. Langage du politique*, 105, 45–66.

CMED (Commission Mondiale sur l'Environnement et le Développement) (1987). *Our Common Future*. Éditions du Fleuve, Montreal.

Coppolecchia, E.K. (2010). The greenwashing deluge: Who will rise above the waters of deceptive advertising? *University Miami Law Review*, 64, 1353–1405.

Delforce, N.J. and Noyer, J. (1999). Pour une approche interdisciplinaire des phénomènes de médiatisation : constructivisme et discursivité sociale. *Études de communication*, 22, 13–40.

Delmas, M.C. and Cuerel Burbano, V. (2011). The drivers of greenwashing. *California Management Review*, 54(1), 64–87.

Funtowicz, S.O. and Ravetz, J.R. (1991). A new scientific methodology for global environmental issues. In *Ecological Economics: The Science and Management of Sustainability*, Costanza, R. (ed.). Columbia University Press, New York, 137–152.

Funtowicz, S.O. and Ravetz, J.R. (1993). Science for the post-normal age. *Futures*, 25(7), 739–755.

Jeanneret, Y. (2010). L'optique du sustainable : territoires médiatisés et savoirs visibles. *Questions de communication*, 17(1), 59–80.

Lapointe, A. (2007). RSE et DD, des pratiques en quête de légitimité. In *Développement durable et communication, au-delà des mots, pour un véritable engagement*, Tremblay, S. (ed.). Presses de l'Université du Québec, Montreal, 64–76.

Le Crosnier, H.N. (2013). Sciences participatives ou ingénierie sociale : quand amateurs et chercheurs co-produisent des savoirs. *Hermès, La Revue*, 67, 68–74.

Loneux, C. (2010). Enjeux de gouvernance à l'épreuve des discours de la RSE. *Communication et organisation*, 37(1), 53–63.

Loneux, C. (2016). La responsabilité sociale des entreprises comme soft law : formes et enjeux de régulation, de dialogue et de frontières. *Revue française des sciences de l'information et de la communication*, 9, 3–17.

Milstein, T. (2011). Nature identification: The power of pointing and naming. *Environmentak Communication: A Journal of Nature and Culture*, 5(1), 3–24.

Neveu, E. (1999). L'approche constructiviste des "problèmes publics". Un aperçu des travaux anglo-saxons. *Études de communication*, 22, 41–57.

Parguel, B.B.M. (2011). How sustainability ratings might deter "greenwashing": A closer look at ethical corporate communication. *Journal of Business Ethics*, 102, 15–28.

Parrenin, F. (2020). Biodiversité et changement climatique : entre discours du spécialiste et discours vulgarisé. *Les Carnets du Cediscor. Publication du Centre de recherches sur la didacticité des discours ordinaires*, 15, 33–46.

Pascual Espuny, C. (2007). Le développement durable : promesse d'un changement paradigmatique ? Étude d'un processus discursif et négocié : un exemple, REACH. PhD Thesis, Paris 4 Sorbonne, Paris.

Pascual Espuny, C. (2015). Le greenwashing, contre-indice de maturité de la communication environnementale ? In *La communication environnementale*, Libaert, T. (ed.). CNRS Edition, Paris, 199–2014.

Raffin, J.-P. (2007). De la protection de la Nature à la gouvernance de la biodiversité. *Écologie et Politique*, 30, 97–109.

TerraChoice (2010). The sins of greenwashing: Home and family edition [Online]. Available at: www.sinsofgreenwashing.org.

Tremblay, S. (ed.) (2007). Du sous-développemet au développement durable, l'émergence d'un concept. In *Développement durable et communication, au-delà des mots, pour un véritable engagement*. Presses de l'Université du Québec, Montreal.

Vos, J. (2014). Actions speak louder than words: Greenwashing in corporate America. *Notre Dame Journal of Law, Ethics and Public Policy*, 23, 673–697.

Conclusion

Communication on Plants Emerging within Contemporary Societies: Beyond the Human World

Given the difficulties in predicting how contemporary relationships with the living world are being and will continue to be reconfigured, we believe that relationships with the plant world, offering more hermetic or less proxemic perspectives than relationships with animals, will help to lend perspective to the reconfiguration of our relationships with the natural environment. Because of its distance, the plant world offers us the opportunity to open up or overhaul the norms, discourse and "glass walls" by which traditional representations and relations are limited.

This is the aim of recently published works, some of which have enjoyed extensive uptake on a global scale, such as Escobar (2018), Hiernaux (2022), Catellani et al. (2023), Johns et al. (2024) and Okon (2024).

In *Sentir-penser avec la terre. L'écologie au-delà de l'Occident*, Escobar (2018), in view of the breakdown of the tired dichotomy between nature and culture, collects other visions of the world and other ways of taking care of it. Thus, he encourages his readers to rethink ecological practices and adopt cosmologies that go beyond the traditional political frameworks of environmentalism, notably based on the diverse range of knowledge and activist practices.

Conclusion written by Amélie COULBAUT-LAZZARINI, Natacha SOUILLARD and Céline PASCUAL ESPUNY.

Johns et al. (2014) contribute to this exploration of practices with their article *Dear Tampa Bay: Creating and leveraging a climate resilience documentary film using story-telling for citizen engagement*. In it, they show the major importance of the storytelling sections of the film on the viewer's ability to memorize the facts, maintain a feeling of urgency and feed the desire to work together to promote local resilience.

Resilience and care are also at the heart of the article by Okon (2024), *Care and carelessness in the Niger Delta: Nature's gift, humanity's ineptness*, where he argues that the media must reposition environmental matters as major concerns that require care and attention. The author explains that it is important, in parallel, to denounce behavior that falls under the category of environmentally harmful practices.

Catellani et al. (2023), in *The intelligence of plants: A new narrative? Notes on a contemporary cultural phenomenon*, argue that the new narrative, which is the intelligence of plants, poses a challenge to western naturalism as defined by Descola, completely upsetting its ontological structure.

The same idea is pursued by Hiernaux (2022) in his article *La communication chez les plantes : biologie et performativité*, in which he interprets the differences and similarities between human language and plant communication, using tools relating to speech act theory.

Also, while the term "plant communication" may, at first glance, seem esoteric or naturalist and not easily compatible with the approaches of human and social sciences, recent scientific literature positions it clearly as part of an emerging dynamic – the very same dynamic, in fact, to which this book belongs. The constituent chapters reflect a shift in the view we take of environmental objects and topics through the lens of information and communication sciences. Indeed, from semiotics to corporate communication, from mediation to discourse analysis, conventional tools and approaches are applied in new ways here, while new ones are also developed.

These scientific approaches reflect the current rise, particularly in French society, of attention to the plant world, which seems to be the object of a process of publicity. Biodiversity, relationships with nature, territories and the ecological transition (etc.) are becoming matters for public debate, as though the construction of collective narratives could run alongside, or even outpace, the changes currently under way or to come. The budgets allocated are also highly noteworthy in a period when budgets devoted to research are being slashed; this can be seen in the 2024 call for projects as part of the ÉRABLE program, driven in France by the Ministry of Culture, on the theme "Creating a biodiversity narrative in the territories".

In this context, the foregoing chapters analyze how the (new) relationships with the plant world cast doubt on public policies cover, and also the need to "write" about these relationships, such as the approach of an anthropologist who compiles environmental narratives, viewed as so many stories or myths; that of an artist, who invents futures; and that of a communicator seeking to make their discourse performative, already present or emerging, but all overshadowed. Is it a matter of creating narratives surrounding the transitions to cause them to come about and/or lending visibility to those that are already in progress?

We speak of overhauling the way in which we view the world and the narratives to which we cleave, but this book also advocates for hybridity, both of objects and perspectives. It makes a contribution to the ongoing construction of a multi-dimensional, transdisciplinary field which, through the lens of the plant world, helps us to rethink the Anthropocene in light of relations between humans and non-humans, and propose plant communication as a sub-field of environmental communication.

Beyond the axes that we have identified, and on which the summary of this book is constructed, focal points emerge and appear to form structural frameworks. Firstly, there is polarity, with an anthropocentric pole (the dimensions of care and utility, environmental service, esthetics and a managerial dimension), stemming from the instrumental value that the plant world holds for humans. There is a second, eco-centric pole, in which human/nonhuman relationships are intrinsically valued and valuable (preservation, coevolution, symbiosis, reliance and sensory input). This polarity shapes spaces of what can be said, which work in terms of diverse links with the plant world. It again casts doubt on the long-held western dichotomy (now largely discredited) whereby the concept of nature is set against that of culture.

Aside from this polarity, throughout the 11 chapters, themes begin to emerge and links are created. There are four particularly salient aspects to which we can point: the sensible or sensory dimension, the relational approach, the notion of mediation and the paradoxical dimension of certain relationships.

This latter dimension is discussed in Chapters 3, 5 and 7, revealed in part by "hybrid objects" as described by Latour (1991), which have both cultural and natural characteristics (Chapter 5); by incentivization to swarm certain places, which can have potentially destructive consequences for places that are known as unspoiled paradises (Chapter 3); and by access to sensory knowledge, which encourages an immersive relationship with nature by means of technological mediation (Chapter 7).

In addition, the relational dimension is at the forefront of seven of the 11 chapters (Chapters 1, 3, 4, 6, 7, 8 and 10). From the question of relational values (Chapter 1) to that of "relational knowledge" (Chapter 8), from intimate and interdependent relationships (Chapter 6) to reliance (Chapter 4), the book reflects the importance of this aspect and its various declinations throughout.

The sensible and sensory dimension is included in over half of the chapters (Chapters 2, 3, 7, 8 and 9), which hold that the sensory dimension should be added to the characteristics of the experiences of encounters with plant life. The concept of "sensification", borrowed from Coccia and taken up by Zengiaro (Chapter 2), brings a veritable ontological upset, which casts doubt on the dichotomy between mind and meaning.

Finally, this book looks towards desirable futures, normative outlooks that are in the process of being reconfigured, with paradoxical situations revealing new narrative frameworks in which the living world, as a whole, is no longer thought of as an object, but as a subject in its own right.

References

Catellani, A., Cholet, C., Pascual Espuny, C. (2023). The intelligence of plants: A new narrative? Notes on a contemporary cultural phenomenon. *Environmental Communication*, 17(6), 618–633. doi: 10.1080/17524032.2023.2234654.

Escobar, A. (2018). *Sentir-penser avec la terre. L'écologie au-delà de l'Occident*. Le Seuil, Paris.

Hiernaux, Q. (2022). La communication chez les plantes : biologie et performativité. *Communications*, 110(1), 57–73. doi: 10.3917/commu.110.0057.

Johns, R., Raines, S., Moore, A., Hill, M., Lansky, P., Pathak, A. (2024). Dear Tampa Bay: Creating and leveraging a climate resilience documentary film using story-telling for citizen engagement. *Environmental Communication*, 18(6), 712–728. doi: 10.1080/17524032.2024.2315181.

Latour, B. (1991). *Nous n'avons jamais été modernes. Essai d'anthropologie symétrique*. La Découverte, Paris.

Okon, G.B. (2024). Care and carelessness in the Niger Delta: Nature's gift, humanity's ineptness. *Environmental Communication*, 18(1–2), 162–166. doi: 10.1080/17524032.2023.2296846.

List of Authors

Joshua T.L. ANDERSON
University of Arizona
Tucson
United States

Pierre BESLAY
CNAM
Université de Reims Champagne
Ardenne
France

Marine BRUN-FRANZETTI
PRISM
Aix-Marseille Université
France

Andrea CATELLANI
ILC-RECOM
Université catholique de Louvain
Belgium

Amélie COULBAUT-LAZZARINI
GRESEC
Université Grenoble Alpes
France

Natacha CYRULNIK
PRISM
Aix-Marseille Université
France

Matthew EASTIN
University of Texas
Austin
United States

Marie-Caroline HEÏD
LERASS
Université de Montpellier
Paul Valéry
France

Philippe HERT
Centre Norbert Elias
CNRS-AMU-UA
Aix-Marseille Université
France

Sarah LABELLE
LERASS
Université de Montpellier
Paul Valéry
France

Megan LAMONICA
University of Texas
Austin
United States

Emma LAURENT
LERASS
Université de Montpellier
Paul Valéry
France

Erica LIPPERT
LaDisco
Université libre de Bruxelles
Brussels
Belgium

Catherine LONEUX
PREFics
Université Rennes 2
France

Valérie MÉLIANI
LERASS
Université de Montpellier
Paul Valéry
France

Céline PASCUAL ESPUNY
IMSIC
Aix-Marseille Université
France

Eva SANDRI
LERASS
Université de Montpellier
Paul Valéry
France

Natacha SOUILLARD
MCF
IMSIC
Aix-Marseille Université
France

Nathanaël WADBLED
Université de Pau et des
Pays de l'Adour
France

Nicolas ZENGIARO
University of Turin
Italy

Index

L, M

O, P

R, S

T, W

Other titles from

ISTE

in

Science, Society and New Technologies

2025

ACCART Jean-Philippe
Marketing for Libraries and Information Services: Systems and Developments

CHRISTOFLE Sylvie
Tourism Agility in Times of Crisis and Uncertainty 1: Between Technology and Society (Tourism and Mobility Systems Set – Volume 11)
Tourism Agility in Times of Crisis and Uncertainty 2: Between Marketing-Communication and Sustainability
(Tourism and Mobility Systems Set – Volume 12)

DEBOS Franck
The Ethics of Artificial Intelligence (Digital Sciences Set – Volume 5)

KAKKAR Misha, HASTEER Nitasha, SINDHWANI Rahul, TURPIN Marita, DAVIM J. Paulo
Strategic Framework and Intelligent Solutions for Sustainable Cities and Communities

LEVEAU Pierre-Henry
Tourism, Technologies and Consumption in the 5.0 Era: Discovery and Reflection in the Face of Organizational and Societal Issues (Tourism and Mobility Systems Set – Volume 10)

MEATCHI Sourou
Revenue Management in the Age of Artificial Intelligence: Towards Ethical and Responsible Price Management in Air Transport, Tourism and Hospitality (Tourism and Mobility Systems Set – Volume 13)

N'GOALA Gilles, PAULHIAC SCHERRER Florence, DURIF Fabien
The Smart and Resilient City: Models and Ways of Life

2024

BARTHES Angela, CIBIEN Catherine, ROMAGNY Bruno
Biosphere Reserves and Sustainable Development Goals 1: Scientific and Practical Educational Issues in the Mediterranean (Territory Development Set – Volume 3)
Biosphere Reserves and Sustainable Development Goals 2: Issues, Tensions, Processes and Governance in the Mediterranean (Territory Development Set – Volume 4)

BOTERO Nataly, LEDOUBLE Hélène, ALLARD-HUVER François
Pesticides: Pluridisciplinary Dialogues in Social and Human Sciences (Communication, Environment, Science and Society Set – Volume 4)

CATELLANI Andrea, LITS Grégoire
Transitions in Tension: Controversies and Tensions Around Ecological Transitions (Communication, Environment, Science and Society Set – Volume 3)

DAVALLON Jean
Heritage Traces in the Making: A Communicational Analysis of Modes of Heritagization (Traces Set – Volume 6)

DEJEAN Maxime, L'HOSTIS Marine, LI Meng
Chinese Outbound Tourism: From Source Markets to Destinations (Tourism and Mobility Systems Set – Volume 8)

GAY Jean-Christophe
Tourismophobia: From "Mass Tourism" to "Overtourism"
(Tourism and Mobility Systems Set – Volume 9)

JULIA Guilhem, FAUCHON Anne, KANAWATI Rushed
AI, Healthcare and Law

KOHLMANN Émilie
Linking with Nature in the Digital Age
(Communication, Environment, Science and Society Set – Volume 1)

LEDOUBLE Hélène
Popularizing Science: The Complex Terminological Interactions between Scientific and Press Discourses within the Field of Agroecology
(Communication, Environment, Science and Society Set – Volume 2)

LOUART Pierre, PIRIOU Jérôme, VIOLIER Philippe
Tourist Destinations According to Stakeholder Strategies: From Clusters to Local Tourist Systems
(Tourism and Mobility Systems Set – Volume 7)

2023

APCHAIN Thomas
Favela Tours: Building Otherness in the Favelas of Rio de Janeiro
(Tourism and Mobility Systems Set – Volume 6)

CARDON Alain, EL HAMI Abdelkhalak
Fundamental Generation Systems: Computer Science and Artificial Consciousness, the Informational Field of Generation of the Universe, the Sixth Sense of Living Beings
(Digital Sciences Set – Volume 4)

CURULLI Irene G., IKIZ KAYA Deniz, KHAEFI Arghavan
Heritage Education for Climate Action
(Research in Architectural Education Set – Volume 2)

ELAMÉ Esoh
The Sustainable City in Africa Facing the Challenge of Liquid Sanitation
(Territory Development Set – Volume 2)

JURCZENKO Emmanuel
Climate Investing: New Strategies and Implementation Challenges

LYUBAREVA Inna, WALDECK Roger
Diversity of Methodological Approaches in Social Sciences: Example of the Analysis of Media and Online Information
(Modeling Methodologies in Social Sciences Set – Volume 3)

PARET Dominique, CRÉGO Pierre, SOLÈRE Pauline
Smart Patches: Biosensors, Graphene and Intra-Body Communications

VENTRE Daniel, GUILLOT Philippe
Electronic Communication Interception Technologies and Issues of Power

VENTRE Daniel, LOISEAU Hugo
Cybercrime during the SARS-CoV-2 Pandemic (2019-2022): Evolutions, Adaptations, Consequences
(Cybersecurity Set – Volume 3)

2022

AIT HADDOU Hassan, TOUBANOS Dimitri, VILLIEN Philippe
Ecological Transition in Education and Research

CARDON Alain
Information Organization of The Universe and Living Things: Generation of Space, Quantum and Molecular Elements, Coactive Generation of Living Organisms and Multiagent Model
(Digital Sciences Set – Volume 3)

CAULI Marie, FAVIER Laurence, JEANNAS Jean-Yves
Digital Dictionary

DAVERNE-BAILLY Carole, WITTORSKI Richard
Research Methodology in Education and Training: Postures, Practices and Forms
(Education Set – Volume 12)

ELAMÉ Esoh
Sustainable Intercultural Urbanism at the Service of the African City of Tomorrow
(Territory Development Set – Volume 1)

FLEURET Sébastien
A Back and Forth Between Tourism and Health: From Medical Tourism to Global Health
(Tourism and Mobility Systems Set – Volume 5)

KAMPELIS Nikos, KOLOKOTSA Denia
Smart Zero-energy Buildings and Communities for Smart Grids
(Engineering, Energy and Architecture Set – Volume 9)

2021

BARDIOT Clarisse
Performing Arts and Digital Humanities: From Traces to Data
(Traces Set – Volume 5)

BENSRHAIR Abdelaziz, BAPIN Thierry
From AI to Autonomous and Connected Vehicles: Advanced Driver-Assistance Systems (ADAS)
(Digital Sciences Set – Volume 2)

BÉRANGER Jérôme
Societal Responsibility of Artificial Intelligence: Towards an Ethical and Eco-responsible AI
(Technological Prospects and Social Applications Set – Volume 4)

CORDELIER Benoit, GALIBERT Oliver
Digital Health Communications
(Technological Prospective and Social Applications Set – Volume 5)

DOUAY Nicolas, MINJA Michael
Urban Planning for Transitions

GALINON-MÉLÉNEC Béatrice
The Trace Odyssey 1: A Journey Beyond Appearances
(Traces Set – Volume 4)

HENRY Antoine
Platform and Collective Intelligence: Digital Ecosystem of Organizations

KOLOKOTSA Denia, KAMPELIS Nikos
Smart Buildings, Smart Communities and Demand Response
(Engineering, Energy and Architecture Set – Volume 8)

LE LAY Stéphane, SAVIGNAC Emmanuelle, LÉNEL Pierre, FRANCES Jean
The Gamification of Society
(Research, Innovative Theories and Methods in SSH Set – Volume 2)

RADI Bouchaïb, EL HAMI Abdelkhalak
Optimizations and Programming: Linear, Non-linear, Dynamic, Stochastic and Applications with Matlab
(Digital Sciences Set – Volume 1)

2020

ALAKTIF Jamila, CALLENS Stéphane
Migration and Climate Change: From the Emergence of Human Cultures to Contemporary Management in Organizations

BARNOUIN Jacques
The World's Construction Mechanism: Trajectories, Imbalances and the Future of Societies
(Interdisciplinarity Between Biological Sciences and Social Sciences Set – Volume 4)

ÇAĞLAR Nur, CURULLI Irene G., SIPAHIOĞLU Işıl Ruhi, MAVROMATIDIS Lazaros E.
Thresholds in Architectural Education
(Engineering, Energy and Architecture Set – Volume 7)

DUBOIS Michel J.F.
Humans in the Making: In the Beginning was Technique
(Social Interdisciplinarity Set – Volume 4)

ETCHEVERRIA Olivier
The Restaurant, A Geographical Approach: From Invention to Gourmet Tourist Destinations
(Tourism and Mobility Systems Set – Volume 3)

GREFE Gwenaëlle, PEYRAT-GUILLARD Dominique
Shapes of Tourism Employment: HRM in the Worlds of Hotels and Air Transport
(Tourism and Mobility Systems Set – Volume 4)

JEANNERET Yves
The Trace Factory
(Traces Set – Volume 3)

KATSAFADOS Petros, MAVROMATIDIS Elias, SPYROU Christos
Numerical Weather Prediction and Data Assimilation
(Engineering, Energy and Architecture Set – Volume 6)

MARTI Caroline
Cultural Mediations of Brands: Unadvertization and Quest for Authority
(Communication Approaches to Commercial Mediation Set – Volume 1)

MAVROMATIDIS Lazaros
Climatic Heterotopias as Spaces of Inclusion: Sew Up the Urban Fabric
(Research in Architectural Education Set – Volume 1)

MOURATIDOU Eleni
Re-presentation Policies of the Fashion Industry: Discourse, Apparatus and Power
(Communication Approaches to Commercial Mediation Set – Volume 2)

SCHMITT Daniel, THÉBAULT Marine, BURCZYKOWSKI Ludovic
Image Beyond the Screen: Projection Mapping

VIOLIER Philippe, with the collaboration of TAUNAY Benjamin
The Tourist Places of the World
(Tourism and Mobility Systems Set – Volume 2)

2019

BRIANÇON Muriel
The Meaning of Otherness in Education: Stakes, Forms, Process, Thoughts and Transfers
(Education Set – Volume 3)

DESCHAMPS Jacqueline
Mediation: A Concept for Information and Communication Sciences
(Concepts to Conceive 21st Century Society Set – Volume 1)

DOUSSET Laurent, PARK Sejin, GUILLE-ESCURET Georges
Kinship, Ecology and History: Renewal of Conjectures
(Interdisciplinarity Between Biological Sciences and Social Sciences Set – Volume 3)

DUPONT Olivier
Power: A Concept for Information and Communication Sciences
(Concepts to Conceive 21st Century Society Set – Volume 2)

FERRARATO Coline
Prospective Philosophy of Software: A Simondonian Study
(Technological Prospects and Social Applications Set – Volume 3)

GUAAYBESS Tourya
The Media in Arab Countries: From Development Theories to Cooperation Policies

HAGÈGE Hélène
Education for Responsibility
(Education Set – Volume 4)

LARDELLIER Pascal
The Ritual Institution of Society
(Traces Set – Volume 2)

LARROCHE Valérie
The Dispositif: A Concept for Information and Communication Sciences
(Concepts to Conceive 21st Century Society Set – Volume 3)

LATERRASSE Jean
Transport and Town Planning: The City in Search of Sustainable Development

LELEU-MERVIEL Sylvie, SCHMITT Daniel, USEILLE Philippe
From UXD to LivXD: Living eXperience Design

LENOIR Virgil Cristian
Ethically Structured Processes
(Innovation and Responsibility Set – Volume 4)

LOPEZ Fanny, PELLEGRINO Margot, COUTARD Olivier
Local Energy Autonomy: Spaces, Scales, Politics
(Urban Engineering Set – Volume 1)

METZGER Jean-Paul
Discourse: A Concept for Information and Communication Sciences
(Concepts to Conceive 21st Century Society Set – Volume 4)

MICHA Irini, VAIOU Dina
Alternative Takes to the City
(Engineering, Energy and Architecture Set – Volume 5)

PÉLISSIER Chrysta
Learner Support in Online Learning Environments

PIETTE Albert
Theoretical Anthropology or How to Observe a Human Being
(Research, Innovative Theories and Methods in SSH Set – Volume 1)

PIRIOU Jérôme
The Tourist Region: A Co-Construction of Tourism Stakeholders
(Tourism and Mobility Systems Set – Volume 1)

PUMAIN Denise
Geographical Modeling: Cities and Territories
(Modeling Methodologies in Social Sciences Set – Volume 2)

WALDECK Roger
Methods and Interdisciplinarity
(Modeling Methodologies in Social Sciences Set – Volume 1)

2018

BARTHES Angela, CHAMPOLLION Pierre, ALPE Yves
Evolutions of the Complex Relationship Between Education and Territories
(Education Set – Volume 1)

BÉRANGER Jérôme
The Algorithmic Code of Ethics: Ethics at the Bedside of the Digital Revolution
(Technological Prospects and Social Applications Set – Volume 2)

DUGUÉ Bernard
Time, Emergences and Communications
(Engineering, Energy and Architecture Set – Volume 4)

GEORGANTOPOULOU Christina G., GEORGANTOPOULOS George A.
Fluid Mechanics in Channel, Pipe and Aerodynamic Design Geometries 1
(Engineering, Energy and Architecture Set – Volume 2)
Fluid Mechanics in Channel, Pipe and Aerodynamic Design Geometries 2
(Engineering, Energy and Architecture Set – Volume 3)

GUILLE-ESCURET Georges
Social Structures and Natural Systems: Is a Scientific Assemblage Workable?
(Social Interdisciplinarity Set – Volume 2)

LARINI Michel, BARTHES Angela
Quantitative and Statistical Data in Education: From Data Collection to Data Processing
(Education Set – Volume 2)

LELEU-MERVIEL Sylvie
Informational Tracking
(Traces Set – Volume 1)

SALGUES Bruno
Society 5.0: Industry of the Future, Technologies, Methods and Tools
(Technological Prospects and Social Applications Set – Volume 1)

TRESTINI Marc
Modeling of Next Generation Digital Learning Environments: Complex Systems Theory

2017

ANICHINI Giulia, CARRARO Flavia, GESLIN Philippe,
GUILLE-ESCURET Georges
Technicity vs Scientificity: Complementarities and Rivalries
(Interdisciplinarity Between Biological Sciences and Social Sciences Set – Volume 2)

DUGUÉ Bernard
Information and the World Stage: From Philosophy to Science, the World of Forms and Communications
(Engineering, Energy and Architecture Set – Volume 1)

GESLIN Philippe
Inside Anthropotechnology: User and Culture Centered Experience
(Social Interdisciplinarity Set – Volume 1)

GORIA Stéphane
Methods and Tools for Creative Competitive Intelligence

KEMBELLEC Gérald, BROUDOUX Evelyne
Reading and Writing Knowledge in Scientific Communities: Digital Humanities and Knowledge Construction

MAESSCHALCK Marc
Reflexive Governance for Research and Innovative Knowledge
(Responsible Research and Innovation Set - Volume 6)

PARK Sejin, GUILLE-ESCURET Georges
Sociobiology vs Socioecology: Consequences of an Unraveling Debate
(Interdisciplinarity Between Biological Sciences and Social Sciences Set – Volume 1)

PELLÉ Sophie
Business, Innovation and Responsibility
(Responsible Research and Innovation Set – Volume 7)

2016

ANDRÉ Michel, SAMARAS Zissis
Energy and Environment
(Research for Innovative Transports Set – Volume 1)

BLANQUART Corinne, CLAUSEN Uwe, JACOB Bernard
Towards Innovative Freight and Logistics
(Research for Innovative Transports Set – Volume 2)

COHEN Simon, YANNIS George
Traffic Management
(Research for Innovative Transports Set – Volume 3)

EL FALLAH SEGHROUCHNI Amal, ISHIKAWA Fuyuki, HÉRAULT Laurent, TOKUDA Hideyuki
Enablers for Smart Cities

GIANNI Robert
Responsibility and Freedom: The Ethical Realm of RRI
(Responsible Research and Innovation Set – Volume 2)

GRUNWALD Armin
The Hermeneutic Side of Responsible Research and Innovation
(Responsible Research and Innovation Set – Volume 5)

LENOIR Virgil Cristian
Ethical Efficiency: Responsibility and Contingency
(Responsible Research and Innovation Set – Volume 1)

PELLÉ Sophie, REBER Bernard
From Ethical Review to Responsible Research and Innovation
(Responsible Research and Innovation Set – Volume 3)

REBER Bernard
Precautionary Principle, Pluralism and Deliberation: Science and Ethics
(Responsible Research and Innovation Set – Volume 4)

TORRENTI Jean-Michel, LA TORRE Francesca
Materials and Infrastructures 1
(Research for Innovative Transports Set – Volume 5A
Materials and Infrastructures 2
(Research for Innovative Transports Set – Volume 5B)

VENTRE Daniel
Information Warfare – 2nd edition

YANNIS George, COHEN Simon
Traffic Safety
(Research for Innovative Transports Set – Volume 4)

2015

BRONNER Gérald
Belief and Misbelief Asymmetry on the Internet

2009

VENTRE Daniel
Information Warfare

Printed and bound by CPI Group (UK) Ltd, Croydon, CR0 4YY

24/08/2026

14947835-0002